RED R
U1896

D0502110

About Island Press

Island Press, a nonprofit organization, publishes, markets, and distributes the most advanced thinking on the conservation of our natural resources—books about soil, land, water, forests, wildlife, and hazardous and toxic wastes. These books are practical tools used by public officials, business and industry leaders, natural resource managers, and concerned citizens working to solve both local and global resource problems.

Founded in 1978, Island Press reorganized in 1984 to meet the increasing demand for substantive books on all resource-related issues. Island Press publishes and distributes under its own imprint and offers these services to other nonprofit organizations.

Support for Island Press is provided by Apple Computers, Inc., The Mary Reynolds Babcock Foundation, The Educational Foundation of America, The Charles Engelhard Foundation, The Ford Foundation, The George Gund Foundation, The William and Flora Hewlett Foundation, The Joyce Foundation, The J. M. Kaplan Fund, The John D. and Catherine T. MacArthur Foundation, The Andrew W. Mellon Foundation, The Joyce Mertz-Gilmore Foundation, The New-Land Foundation, Northwest Area Foundation, The Jessie Smith Noyes Foundation, The J. N. Pew, Jr. Charitable Trust, The Rockefeller Brothers Fund, The Florence and John Schumann Foundation, The Tides Foundation, and individual donors.

About American Rivers

American Rivers, Inc., is a nonprofit conservation organization dedicated to the preservation of the nation's outstanding rivers and their landscapes. Since its founding in 1973, American Rivers has helped preserve 8,000 miles of prime natural river, protect seven million acres of adjacent lands, and stop scores of ecologically destructive dams.

The conservation activities of American Rivers are made up of three separate programs. The hydropower program focuses on directing hydropower development toward appropriate sites and away from outstanding free-flowing rivers, where it does not belong. The federal river protection program focuses on fulfillment of the federal wild and scenic rivers system. Finally, the state rivers program encourages states to initiate statewide rivers inventories and assessments, and then works to obtain state-level protection for outstanding rivers.

For further information about American Rivers, write to 801 Pennsylvania Avenue SE, Washington, D.C. 20003, or telephone (202) 547–6900.

Rivers at Risk

Rivers at Risk

The Concerned Citizen's Guide to Hydropower

John D. Echeverria
Pope Barrow
Richard Roos-Collins

American Rivers

ISLAND PRESS

Washington, D.C. Covelo, California

Text Design by Irving Perkins Associates
Cover Design by Ben Santora
Cover Photograph by Tim Palmer

Library of Congress Cataloging-in-Publication Data

Echeverria, John D.
Rivers at risk : the concerned citizen's guide to hydropower / John D. Echeverria, Pope Barrow, Richard Roos-Collins.
p. cm.
Includes bibliographical references.
ISBN 0-933280-83-1 — ISBN 0-933280-82-3 (pbk.)
1. Hydroelectric power plants—Environmental aspects. 2. United States. Federal Energy Regulatory Commission. 3. Stream conservation—United States—Citizen participation. I. Barrow, Pope, 1942– . II. Roos-Collins, Richard, 1953– III. Title.
TD195.E4E22 1989
333.91′4′0973—dc20 89-19831
CIP

Printed on recycled, acid-free paper

Manufactured in the United States of America

10 9 8 7 6 5 4 3 2 1

Contents

Foreword xi

Acknowledgments xiii

Introduction 1

1. The Center of the Action: The Federal Energy Regulatory Commission 15

The History of Hydro Regulation 16
The Scope of FERC's Jurisdiction 17
FERC's Organization 19
The Courtlike Character of FERC 21
A Word about PURPA 24

2. An Overview of the Regulatory Process 27

The Steps in Original Licensing 28
Relicensing 32
How the Application Review Process Works 33

3. New Projects: Standards and Procedures 43

FERC's Licensing Standards 44
FERC's Responsibility to Consult with Other Entities 52
FERC's Responsibility to Study Environmental Impacts 56
Environmental Standards for PURPA Benefits 59

4. Relicensing: Standards and Procedures 63

An Overview 64
FERC's Relicensing Standards 67

Procedural Aspects of Intervening in Relicensing Proceedings 70
Alternatives to Issuance of a New Power License 71

5. Participating in the FERC Process 79

Learning about a Project 80
Intervening in a FERC Proceeding 83
Rectifying a FERC Error 89
Vigilance after Project Construction 92

6. Raising Issues before FERC 95

General Rules for Making Your Case 96
Potential Issues to Raise 98

7. Other Tools for Protecting Rivers 111

Federal Wild and Scenic Rivers System 112
Federal Legislative Protection 115
The Clean Water Act 116
Federal Land Management Agencies 118
State River Protection 119
The Northwest Power Planning Council 121

8. Strategies for Effectively Dealing with Hydroelectric Projects 123

Strategic Planning 124
Finding Allies 125
Legal and Technical Expert Assistance 126
Raising Money 128
Favorable Press Coverage 129

Appendix A. FERC Addresses and Telephone Numbers 133

Appendix B. Conservation Groups Concerned About Hydropower Issues 134

Appendix C. Model Motion to Intervene 135

Appendix D. A FERC Licensing Order 148

Appendix E. FERC's NEPA Regulations 159

Appendix F. FERC's Freedom of Information Regulations 169

Appendix G. FERC's 1989–1999 Relicensing Workload 183

Appendix H. How to Find and Use the Legal Documents on Which FERC Relies 197

Glossary of Hydropower Terms 199

Index 203

Foreword

A quirk in a law Congress passed a decade ago has spawned a new class of entrepreneurs who are taking advantage of an opportunity to generate minuscule quantities of hydroelectric power—and to thereby create private "cash registers" for themselves—by obtaining licenses from the Federal Energy Regulatory Commission (FERC) to install dams on hundreds of the nation's remaining free-flowing rivers. This quiet conservation crisis is upon us: applications already docketed at FERC make it certain that as many decisions will be made about rivers in the next few years as ever before in our nation's history.

Between now and 1993, more than 200 existing hydroelectric projects will come up for relicensing, during which their economic and environmental impacts must be reevaluated. This process gives river conservationists a rare opportunity to mitigate past damage, to restore a river to its natural state, or to require improvements in existing dam facilities.

Many of these projects have provoked controversies between private developers, with their economic interests, and communities or conservationists with their campaigns to save natural rivers as amenities for cities or as irreplaceable sites for water-based forms of outdoor recreation.

This is a how-to-save-your-local-river book, and it is appearing just in the nick of time. It contains legal tools and insights concerning pertinent environmental laws that will enable river savers to kill, or modify, bad hydro plans presented to FERC. And it outlines successful strategies that have been used to delay or defeat ill-conceived projects.

The fight to preserve America's remaining wild and scenic streams will be one of the most important conservation struggles of the 1990s, and I am confident this book will be a useful guide for citizens and communities embroiled in these contests.

Stewart L. Udall
Phoenix

Acknowledgments

Rivers at Risk represents a collaborative effort to distill the knowledge, the wisdom, and the sometimes painful experiences of numerous individuals and organizations that have struggled to preserve free-flowing rivers from hydropower development. Our list of acknowledgments is incomplete. Indeed, some of those whose insights and suggestions were most helpful prefer that we not acknowledge their contributions at all. Persons who have provided valuable assistance in the preparation of *Rivers at Risk,* in addition to our forbearing wives and children, include David Conrad of the National Wildlife Federation, J. Glenn Eugster of the National Park Service, Harriet LaFlamme, J. V. Henry of California Save Our Streams, Dawn King, Ronald Kreisman of the Natural Resources Council of Maine, Chuck Magraw of Wilson and Cotter, Suzi Wilkins of American Rivers, Scott Reed, Pete Skinner of the American Whitewater Affiliation, Elizabeth Andrews and Ron Stork of Friends of the River, and the staff of the FERC Office of Hydropower Licensing.

Rivers at Risk

Introduction

SIX HUNDRED THOUSAND MILES of what had been free-flowing rivers now lie quiet behind concrete. They were once recognizable as gold-medal trout fisheries, adventure-class white-water runs, irreplaceable habitats of endangered aquatic and stream-bank biota, important historic and archaeological sites, unique geologic features, scenic climaxes, and outdoor playgrounds for all manner of recreationists. Now they are motionless lagoons.

No one knows precisely how many dams plug the nation's rivers. The Environmental Protection Agency estimates at least 68,000, the National Park Service puts the total at 75,000, and the *Wall Street Journal* recently said "80,000 or so." The U.S. Soil Conservation Service notes that if you count farm ponds, the number of impoundments soars to 2,000,000. The figures include at least 750 Army Corps of Engineers or Bureau of Reclamation dams, more than 1,700 dams on Bureau of Land Management lands, 300 on Indian lands, 170 in National Wildlife Refuges, 260 in the National Park System, and 50,000 general purpose dams.

By early 1988, according to the Federal Energy Regulatory Commission (FERC), more than 2,000 hydro projects were operating. A project often consists of several dams, so the number of actual structures is even larger. The Environmental Protection Agency has estimated private hydro dams at 15,000. The large discrepancy in the figures suggests the lack of comprehensive oversight given to river development.

So far, America has dammed about 17 percent of its 3.5-million-mile complement of natural rivers, mostly in the last hundred years. By contrast, we have given ironclad protection to about one-fourth of a percent, or 9,000 miles. For each river mile preserved, 65 miles have been dammed. Hydropower dams, the focus of *Rivers at Risk*, are a major contributor to the extinction of rivers.

Between 1984 and 1988, the number of operating projects increased by almost 50 percent. By early 1988, more than 850 development proposals were pending. FERC estimates that 1,500 new dams may eventually be built. Natural rivers will continue to be extinguished.

Rivers at Risk is for anyone trying to deal with hydro dams: the angler concerned for the prospective loss of a favored trout stream; the kayaker, canoeist, or rafter seeking to keep white water white; the biologist restoring anadromous salmon to ancestral spawning grounds; the ecologist bent on preserving riparian habitat or rare species and natural communities; the Indian tribe seeking to retain ancient fishing rights; the farmer trying to prevent a field from being confiscated and inundated; the hiker trying to walk from road to riverside without trespassing on a dam owner's premises; the town council planning literally to turn a city around so its buildings can face its centerpiece river; or the homeowner wanting to sell streamside property. The ordinary people touched most directly by hydropower have been the least equipped to do anything about protecting their birthright of rivers.

WHY WE SHOULD PRESERVE NATURAL RIVERS

The rivers of America serve as the veins of our continent. They supply drinking water, carry vital nutrients toward estuaries, and function as life corridors for aquatic and terrestrial species, linking otherwise isolated habitats upstream and down. They nourish the bottom lands with new, ever richer stores of fertile material and carve, shape, and build the physical environment around them. Like the circulatory system of the human body, our river systems provide their services at no cost to the beneficiaries, enhancing the ecological health and durable beauty of the American landscape.

Though the preservation of natural rivers is first an ecological necessity, the benefits apply to an array of human needs, too. Communities are discovering that rivers kept clean and damless are positive economic influences, especially in metropolitan areas, enhancing local property values and giving habitable character to municipal settings. Rivers provide simple recreations, from self-propelled leisures such as fishing, canoeing, rafting, swimming, hunting, and streamside hiking, to motorized pursuits like boating. Increasingly, river lands are gaining protection because floodplains left free of construction act as giant sponges during flood, giving low-cost, damage-free control of inundation.

Rivers are refuges for the soul, places of spiritual refreshment where the natural flow and play of running water mirror the movement of life itself. They provide for elemental, relatively unadorned experiences in which humankind and nature can come together. Ecologically and aesthetically, rivers are indivisible from the larger American land. When preserved, rivers serve as visible symbols of the care we take as temporary inhabitants and full-time stewards of a living, profoundly beautiful heritage of nature.

Henry David Thoreau said, "Man is rich in proportion to the number of things he can afford to let alone." That succinctly describes the social purpose of preservation. Although people who work toward keeping rivers entire do so partly to stay the prospect of a riverless future, the deeper motive is to improve the quality of life for all. Many conservationists believe that a rich life depends in part on people's ability to experience eternal nature—in the case of a river, to flow *with* it in sympathetic appreciation, whether contemplatively or in literal contact with moving water. Brought close to things natural, people can know the immense silences of the few remaining places on our continent where wildness is still ascendant, can discover for themselves what Emerson called "an original relation with the universe."

But the very nature of a river militates against protecting it easily. A stream is a fluid medium. A reach of river cannot be preserved merely by drawing an official property line around it, for in no comprehensive way is it secure, even if the riparian acreage is formally protected. Upstream pollution can ride in the water, and ambient influences like acid rain can still harm it no matter what. (However, it is equally true that a river can naturally flush itself of many unwanted substances, given world enough and time.) Riverine ecosystems are simply more dynamic than land, harder to "set aside."

And is a stream truly preserved if its waters are forever pristine but its banks are cluttered with slums? Is a river merely a constellation of slippery molecules in a trench? Or is it, as Hal Borland wrote, "the summation of a whole valley"? Something that eludes plain definition resists being staked down by written codes. Perhaps Heraclitus pegged it best: "You cannot step into the same river twice."

Given the protean character of rivers generally, their vulnerability to legally authorized incursion and dewatering, their qualities as natural dividers or straddlers of private properties and political jurisdictions, their attractiveness for streamside development, and the lack of comprehensive protection tools (even for watercourses within managed public lands), it is little wonder that our riverine legacy has suffered so much. Rivers in America seem to belong to everyone and no one at the same time.

America leans perceptibly toward developing any suitable river. Any unpreserved river is a river at risk, and in that context, the question of building yet more dams of any kind must be reevaluated.

HOW DAMS HARM RIVERS

When a concrete and steel plug is inserted, a river loses many natural qualities. So long as the dam remains, the river is never the same again, even if the changes are offset by compensatory actions such as aerating the water, reseeding the stream banks, restocking the fish, or releasing scouring flows. And no mitigation can possibly reconnect segments of river that a dam has severed.

Because of the laws governing hydropower, most dams—often repaired, reconstructed, or expanded over time—are forever. Thus, except for cases of extreme damage from pollution, the least reversible form of river alteration comes from dams.

Streams remain healthy in proportion to how much oxygen they carry. When a river is impounded and robbed of oxygen, the water issuing from the dam is likely to be low in oxygen too. Species that depend on high-quality waters can be completely extirpated from the stream.

As a river deepens and widens into an impoundment behind a new dam, the water column may stratify by temperature. Little oxygen or light reaches the lower strata, and the upper levels develop into a mostly warm water body. The warmth, in turn, is inhospitable to certain cold-water fishes that predominate in damless waters. Some of those species, such as native trout, eventually vanish from the impoundment. The habitat is sometimes taken over by species less valued by anglers.

Reservoirs alter streamside ecology as well. Impounded water drowns some vegetation completely, leaving other plants to survive or die off in a lakeside perimeter resembling a bathtub ring. When the dam operator stores water, the ring is submerged. It emerges when flows are released to spin turbines or irrigate. Extensive mud flats can develop. The broad surface of impounded water invites evaporation, concentrating salts in the pool and diminishing downriver flows. Sediments begin to fill the reservoir, reducing its storage capacity. From the day it is built, a dam is headed toward obsolescence.

The flows coming from a dam, plus dam leakage, rarely meet the hydrologic standards of nature itself. Although a dam operator may be required to release water regularly into the riverbed below a dam, the flow is usually insufficient to maintain the predam ecology. The spurts are too feeble, and they sometimes contain third-rate water from oxygen-poor depths of the pool. Such releases are less voluminous than in nature's own diurnal and seasonal hydrology.

Some fish species are accustomed to natural sedimentation in a stream, and their reproductive and feeding cycles function accordingly. These fish suffer when a dam traps the sediment that otherwise would have gone downstream, while others thrive in the clear flow. However, the water pulsing from a dam can also erode stream-bank materials that act as filters and protect certain other fish species from unwanted sediments. Depending on the size, composition, and timing of the unexpected loads, the eroded sediments threaten to block fish gills, fill up spawning sites, and smother live food.

At some hydro dams, water is funneled into pipes or canals and returned to the river at a downstream powerhouse. This can dewater the riverbed for miles, turning formerly submerged channels into barren, high-relief rock gardens. In the bypassed section of river, it's as if a spigot were turned off. Anglers, canoeists, kayakers, boaters, rafters, and native animals and plants get little use from such stretches.

Because a dam eliminates regular, reliable surge, the riverscape becomes less riverine. Healthy river corridors commonly receive ecological and geologic nourishment through periodic inundation, abrasion, and deposition. A naturally engorged river tears away at the confining geography, removes invading plants, supplies desirable nutrients to the riparian land, and deposits sand and cobbles. A heavily manipulated river, however, can cart away beaches by the ton. Denied the benign rough-and-tumble of real floods, a river suffers qualitatively, undergoing a physical, biologic, and geologic reduction in natural conditions.

The human-caused simplification of nature is an especially critical problem in biotic preservation. River environments serve as last strongholds for certain vanishing ecosystems, species, and natural communities. Many original floodplain forests, for example, have been developed or cultivated to near nonexistence. Scientists believe that native riparian vegetation in floodplains of the lower 48 has been reduced overall by 90 percent. Today, free-flowing rivers and their remnant biology might properly be compared to museums of natural history.

Unfortunately, the value of rivers as refuges and as corridors for genetic transmission does not easily yield to traditional cost-benefit formulation. No matter how spurious the numbers, when we are confronted with short-term calculations that purportedly prove the economic viability of dams, we seem not to value adequately even the vestigial naturalness of rivers. If nothing else, shouldn't we treat them at least as rarities?

Three illustrations show how hydroelectric development can harm rivers.

- *Sayles Flat* is a half-constructed hydro project on the South Fork of the American River in California's Sierra Nevada foothills. The diversion dam and a mile-long penstock would destroy four dramatic cascades. Boating and other recreation would be eliminated along a mile of dewatered stream.

  FERC issued the license in 1983, over conservationists' objections. The developer elected to proceed. However, in 1988, Harriet LaFlamme, a schoolteacher and avid environmentalist, prevailed in a federal appeals court. It ruled that FERC had done inadequate environmental review and suspended the license. The Commission staff recently prepared an environmental assessment proposing that the project be completed, but subject to tougher environmental standards than it originally required. It is uncertain whether the project will be completed.

- At the *Hawks Nest* project on the New River in West Virginia, the powerhouse is downstream of the dam, and the two are connected by a large tunnel and penstock. Although the average annual flow in the New River is a substantial 9,000 cubic feet per second, for many years the five-mile-long bypassed segment—known as The Drys—contained only 25 cubic feet per second. The reductions in flow and the radical variations in water temperature have seriously degraded the fishery. The intake pipe is not adequately screened, so the turbines destroy large numbers of fish.

  In the recent relicensing proceeding for Hawks Nest, the U.S. Fish and Wildlife Service recommended that minimum flows in The Drys be increased to between 2,000 and 4,000 cubic feet per second. Unfortunately, the Federal Energy Regulatory Commission rejected this recommendation and adopted a mere 100-cubic-feet-per-second minimum flow. Conservationists are now appealing the decision.

- The *Columbia Falls* project on Maine's Pleasant River threatens one of the six native salmon populations in America. Constructed in 1981, this tiny project was designed to provide electricity to a few hundred homes. But FERC failed to detect an engineering miscalculation. The project has never operated to its design capacity. The owners now have serious financial problems. Furthermore, improperly designed fishways are preventing Atlantic salmon from migrating, threatening their very existence in the river.

Today, federal and state officials are convinced that the dam never should have been constructed. Plans are underway to tear down the dam and restore the Atlantic salmon's pathway upriver.

As a nation, we are starting to understand that dams can harm a river. We are also beginning to appreciate the cumulative withering-away of America's total riverine heritage, a legacy that has both biologic and cultural dimensions and whose loss therefore represents a significant social deprivation. But the general public's ability to comprehend that dams also destroy economic opportunity has not yet matured, despite a growing number of studies affirming the beneficial economics of river preservation.

We need to begin to value not only the gross national product, but also the gross rivers product—the total value of goods and services provided by free-flowing rivers. These include clean water, pleasant living space, fresh food, natural flood control, stunning or serene beauty, and uncrowded places for quietly taking stock of and giving thanks for one's place in the world. We are not yet fully aware of the positive economic impacts of outfitting, fishing, boating, rafting, or any other business reliant on the intactness, as opposed to the conversion, of the resource that supports it. Maximizing the gross rivers product by limiting river destruction will require a populace better educated to the economics of preservation, better informed about hydropower.

WHAT THE PUBLIC DOESN'T KNOW ABOUT HYDROPOWER

To deal with hydro, you must know something about the Federal Energy Regulatory Commission (FERC), the independent federal agency that regulates nonfederal hydroelectric dams, and about the Federal Power Act, the basic law governing FERC review of hydropower proposals. FERC is a courtlike body, not a natural resource manager or an environmental agency. It responds to formal filings, motions, applications, and interventions, not to straightforward requests. But even ordinary citizens can make a difference in the way FERC deals with rivers. *Rivers at Risk* is meant to help.

The Federal Energy Regulatory Commission, formerly known as the Federal Power Commission, has been judging hydroelectric proposals since 1920. For many decades, the Commission had to consider only a few proposals each year. Since the late 1970s, however, the agency has processed thousands. It now authorizes about 250 new hydro projects annually.

A driving force behind the hydro rush was the Public Utility Regulatory Policies Act of 1978, or PURPA. Its original purpose was to accelerate development of "clean" energy—power from the sun, wind, and water. In enacting PURPA, Congress was responding to a national concern about dependence on foreign oil, reliance on atomic energy, and the specter of a landscape torn asunder in the quest for coal and oil shale. Unfortunately, the law did not simultaneously direct the developers away from preservation-worthy rivers. Evidently, no one had thought out the consequences of a national policy that set no rivers absolutely off limits to hydro (except, of course, the tiny quotient of federally designated "wild and scenic" rivers).

PURPA gave independent power producers the chance to enter the energy business. The law requires utilities to buy electricity from hydro dam operators. The utilities must purchase the power whether or not they need it, usually at rates far exceeding the costs of alternative power. Many hydro projects that would fail outright in the free market have flourished thanks to this protectionism.

Also, in the late 1970s and early 1980s, Congress passed laws that gave developers accelerated depreciation and enormous tax credits. The cost to taxpayers of these federal subsidies has amounted to hundreds of millions of dollars. Some conservationists supported the subsidy concept, but they originally understood that PURPA benefits would apply to hydro projects at existing dams only. FERC nonetheless made PURPA benefits available to new dams and to "old dam" projects at sites where nature had long ago obliterated any sign of a dam. The result was a frenzy of construction.

FERC thinks of itself as a hydropower promoter. During the agency's first 60 years, it only once turned down a project in order to protect recreational and aesthetic values of a river. The Commission viewed the recent hydro-spurring legislation as an endorsement of the agency's prodevelopment policy.

Between 1978 and 1985, FERC received 6,500 dam proposals. Many were uneconomic, but construction nonetheless increased dramatically. Each application for a hydro license, a sort of certificate of partial ownership of a river, represents yet another potential extinction of untrammeled waters. Especially threatened are adventure-class white water (now reduced to less than a percent of the remaining undammed river mileage of the United States), cold-water fisheries, and America's "home" rivers—the local streams, brooks, and rivers that give habitable character to our communities.

In 1986, Congress began to redress the imbalance between hydro development and river protection by adopting the Electric Consumers Protection Act (ECPA), which amended the Federal Power Act. The Commission now must give "equal consideration" to power development and preservation of recreational, ecological, and other values of natural rivers. Congress placed a temporary halt on PURPA benefits for new dam projects and established higher standards for approving future projects. Resource agencies, such as the U.S. Fish and Wildlife Service, were given a stronger hand in setting license conditions.

Nonetheless, the basics of the old rules remain the same. Thus, the impact of hydropower on our dwindling heritage of natural rivers will continue to be extreme. For example, state-"protected" rivers are still not legally immune to FERC's preemptive authority to permit a dam. In energy matters, federal law is supreme. Also, many rivers in national forests, recreation areas, monuments, and wildlife refuges are fair game, especially since developers regard such publicly owned lands as free. The power of federal land managers to stop unwanted hydro development is insignificant. Among agencies, FERC is *capo di capo.*

A startling fact is that developing all the best remaining "small hydro" sites would contribute little to the nation's projected energy requirements. (Mammoth hydro dams are another issue.) If developers were allowed to exploit to full generating capacity every economically viable hydro location likely to pass threshold environmental review, the total output of electricity would equal that of a single, medium-sized coal-fired plant.

That's precious little energy, enough to displace a mere three one-hundredths of a percent of present U.S. fossil fuel usage. Another way to picture it: America could save that much energy by improving automobile efficiency standards by *one-tenth of a gallon,* from 26.5 miles per gallon to 26.6! The thoughtful question is: how many additional stretches of living river is the nation willing to sacrifice for so meager a payoff?

Still, hydro has its advocates, including some conservationists. Some people believe in hydropower because it does not foul the air or conspicuously degrade water. You don't strip-mine to produce it, nor do you irrevocably consume the principal resource itself, which you do in the case of fossil fuels. There are no invisible radioactive emissions, no possibility of meltdown, no poisonous wastes with half-lives of 75,000 years. So viewed, hydropower represents a substantial public good. So founded, the ethics of hydro advocates are sound, no matter how critically we river savers may judge their objectives.

However, other than some legislators, agency officials, and the developers themselves, few hydrophiles know about the artificial market (which *should* be a serious national concern, especially at times of federal deficit), the destructive effects dams have on riverine ecology, the minuscule output of hydropower when judged against projected national energy needs, or the fact that developers face few legal restrictions as to which rivers they can extinguish. The general public itself uncritically embraces hydro as a seemingly chaste means of energy production. Virtually no one seems to acknowledge the loss of irreplaceable natural assets under the ironic guise of "renewable" energy.

Popular belief says the era of big dams is over, a dubious contention. Regardless, the era of the small hydropower dam *is* here and will not pass quickly. America has to reckon now with the full significance of 1,500 proposed new hydro dams. The reckoning will be difficult, partly because our national concept of rivers is so grandiose—we generally judge a river's worthiness by its size and name value. Ironically, most of us will never see the big rivers in their wild settings, the Yukons, Colorados, Allagashes, upper Missouris, Salmons, Snakes, and others.

For most of us, the price of continued damming will be exacted in the loss of pieces of local culture—the old swimming hole, the childhood fishing site, the favorite streamside picnic area. As these places continue to disappear, perhaps we will respond to the immediate cultural deprivation more deeply and directly than we have, so far, to the fates of remote megarivers, many of which were heavily dammed to no real purpose.

Our nation needs an expanded common definition for *river.* It would honor brand name rivers, and wildness, and ecological and scenic vitality, of course, but also encompass the humane uses of ordinary rivers. The new definition would stipulate that well-made agricultural, rural, and even municipal landscapes comprise preservation-worthy attributes of community rivers. It would classify watercourses by qualities instead of merely by size. It would ask that we think of flowing water, the topography that confines it, the living things that depend on it, and the local cultural associations that enrich it, as a more comprehensive vision of *river.* But until we all are better informed about riverness—that's a long way off—we can expect hydroelectricity to be promoted as part of the mix of "clean" energies of the future, even by those who stand to lose the most, the people closest to nameless rivers.

Meanwhile, the immediate challenges to river conservationists are to kill bad hydro proposals, desubsidize the new-dam market, and limit acceptable hydro development to places where it will cause the least ecological and aesthetic damage, the least disruption in competent human use of a riverscape.

DAM OWNERS DO NOT OWN RIVERS—HYDRO RELICENSING

Before passing the Federal Power Act in 1920, Congress debated the role that private enterprise should have in developing rivers. Some legislators favored using mostly private capital. Others argued that since rivers are public resources, publicly owned enterprises were preferable. The Federal Power Act embodies a compromise between the two positions.

Private companies can build hydro dams, but they cannot "buy" a river. If power is to be harnessed, the developer must obtain a license. That is, government grants the limited right to alter the river for the recognized public good of producing electricity. However, since the definition of *public good* may change over the years, licenses have a maximum term of 50 years. When the license expires, the commission must decide afresh how the resource should be managed, and by whom.

Now is the time to take that fresh look. Between now and 1993, more than 200 existing hydroelectric projects—representing perhaps twice that many dams—will come up for FERC's review. One hundred seventy-five licenses will expire in 1993 alone.

Relicensing represents a singular opportunity to reclaim river values previously lost to hydropower. *Rivers at Risk* tells you how to get the job done. You can obtain improved water releases to enhance recreation, wildlife, fisheries, and other natural values. You can ask FERC to require a dam owner to install adequate upstream and downstream fish passage facilities. You can call for guaranteed public access to dam sites. Relicensing can give you the legal force to make the owner rehabilitate a poorly constructed or unsafe dam.

Relicensing is not an antihydropower process. It recognizes that decisions made years ago may not be adequate today. It is a way to raise a river's recreational usefulness and create a higher degree of naturalness without necessarily causing a significant loss of electricity production.

The Federal Power Act also grants to any person—that is, to a municipality, a state, or a conservation organization like American Rivers or the American Whitewater Affiliation—the right to apply for ownership of existing dams due for relicensing and to convert them to nonpower use. Lest anyone think wresting a dam from its longtime owner is confiscatory, the law makes clear that no taking is involved. The dam owner is assumed to have amortized the capital investment and to have received an adequate return over the term of the original license.

A few dams already have been targeted for possible removal through the relicensing process. Undoubtedly there are more good candidates. American Rivers, in cooperation with state and local groups, is pursuing the removal of Edwards Dam, on the Kennebec River in Augusta, Maine's capital. The century-old dam generates little power, blocks anadromous fish runs, and creates a slack lagoon for 15 miles upriver.

With every successful conservationist intervention in a dam relicensing, a precedent for subsequent relicensings could be set. Whether conservationists try to upgrade dams and therefore upgrade a working river—one likely always to remain harnessed for power production—or try to eliminate harmful or unnecessary dams, the objective is to simulate a natural river. Said differently, the call for controlling hydropower is no more than a modest reassertion of the public's right to balanced management of the public's water.

THE PURPOSE OF THIS BOOK

Rivers at Risk: The Concerned Citizen's Guide to Hydropower is dedicated to two simple propositions: the natural values of many working rivers deserve enhancement; and many damless rivers deserve to remain undammed forever.

John Echeverria, Pope Barrow, and Richard Roos-Collins, attorneys and expert river savers all, have converted into plain prose some of the complex rules governing hydropower. *Rivers at Risk* is for nonspecialists and professionals alike who want to stem future river extinctions. It is also for hydro developers wishing to minimize conflicts with conservationists.

The book aims to help you to (1) intervene in hydropower proceedings; (2) direct hydro development away from rivers deserving preservation; (3) deal efficiently with the Federal Energy Regulatory Commission; and (4) use hydro licensing and relicensing procedures to guarantee adequate fish passage, sufficient water releases, and better public access to rivers.

Rivers at Risk deals with nonfederal hydro dams, which are under the jurisdiction of FERC, rather than with public dams, which are operated by various federal agencies such as the Army Corps of Engineers and the Bureau of Reclamation. Nonfederal projects include those operated by utilities, independent power producers, manufacturers, cities and towns, and so on. These make up 90 percent of existing hydro projects.

American Rivers hopes this book contributes to improving old hydro dams and preventing unneeded new ones, but also to affirmatively protecting many more watercourses. Natural rivers are an endangered part of the American landscape and a beautiful part of our social and ecological inheritance. As Stewart Udall has said elsewhere, America needs to enlarge its heritage of preserved rivers.

W. Kent Olson
President, American Rivers

Chapter 1

The Center of the Action: The Federal Energy Regulatory Commission

FERC's origin and traditional attitudes toward hydro development • the scope of FERC's jurisdiction • FERC's organization • the quasi-judicial character of FERC • PURPA—the Public Utility Regulatory Policies Act

CITIZENS ARE COMMONLY SURPRISED to learn that the fate of their favorite stream may be decided by a little-known agency in Washington, D.C., named the Federal Energy Regulatory Commission—more commonly called FERC or the Commission. Under federal law, the Commission has primary authority to decide whether a proposed hydroelectric project may be built, and on what terms. Various federal and state agencies also play important roles in regulating hydroelectric development, but the Federal Energy Regulatory Commission is at the center of the action.

This chapter provides an overview of the Commission's place in the hydroelectric regulatory process and of its jurisdiction and methods of operation.

THE HISTORY OF HYDROPOWER REGULATION

The Commission, originally known as the Federal Power Commission, was established by the Federal Water Power Act of 1920,[1] renamed the Federal Power Act in 1935.[2] In the early part of this century, Congress became frustrated with the division of authority over hydroelectric development among different agencies and enacted the Federal Water Power Act. Its primary purpose was to reduce administrative confusion by centralizing the planning and regulation of hydroelectric power in a single agency.

Congress's move to streamline the administrative process in 1920 was, in part, to encourage development of river resources for more power. In 1920, hydropower was generally viewed as a boon to economic development, and free-flowing rivers were considered just one more element of the limitless bounty of nature. Today, nearly 70 years later, the public has a more balanced view of hydropower development, and a greater appreciation of the value of free-flowing rivers.

Over the last several decades, in response to these new attitudes, the concentration of authority in the Commission has been steadily eroded, and there has been a corresponding increase in the authority of other federal and state agencies. As a result of the 1972 amendments to the Clean Water Act, the states and the U.S. Army Corps of Engineers now have independent authority to decide whether or not to permit the construction of nonfederal hydroelectric projects. Similarly, the authority of the federal land management agencies to regulate nonfederal hydroelectric development on their lands has, arguably, been strengthened by other reforms to federal laws. (See chapter 7 for a detailed discussion of these laws.)

More recently, Congress enacted the Electric Consumers Protection Act of 1986 (ECPA),[3] adopting important amendments to the Federal Power Act. ECPA strengthens the authority of federal and state fish and wildlife agencies over hydroelectric development. ECPA also directs the Commission to give "equal consideration" to power generation and the recreational and ecological values of free-flowing streams in making its decisions. ECPA requires the Commission, for the first time, to consider whether proposed developments are consistent with state scenic rivers programs and other state or federal river plans.

Despite these important legislative reforms, the Commission has responded slowly to its evolving mandate; the original goal of facilitating hydroelectric development through centralized regulatory control remains entrenched in the thinking at FERC. An understanding of the Commission's institutional bias for promoting hydropower is essential if an effective strategy for dealing with a hydropower proposal is to be developed. The Commission can be convinced, and forced through court action if necessary, to do well by the environment. But citizens may be disappointed if they rely on the goodwill of the Commission to protect the best local boating or fishing stream from inappropriate hydroelectric development.

Though FERC remains at the center of the action, the increasing involvement of other federal and state agencies in the regulatory process means that you should not overlook the possibility of achieving your objectives through these other agencies. Frequently, state and federal resource management agencies are more sympathetic to environmental concerns than is the Commission because these agencies are specifically charged with protecting the resources. Sometimes these agencies have the authority to stop projects, or at least to ensure that adverse environmental effects are mitigated.

THE SCOPE OF FERC'S JURISDICTION

The jurisdiction of the Commission over hydroelectric development was created by the Federal Power Act. In practice, however, the Commission has jurisdiction over the overwhelming majority of nonfederal hydroelectric projects built or proposed in the United States. Thus, most citizens concerned with particular projects will not be affected by the issue of Commission jurisdiction. Nevertheless, for those interested, it is summarized here. The Commission has mandatory jurisdiction over any proposed or existing hydroelectric project[4]

- if it occupies federal public lands or federal reservations, including Forest Service lands, Bureau of Land Management lands, and Indian reservations
- if it is located on a navigable stream, including, for example, any river historically used to transport logs
- if it uses "surplus" water or water power from a federal government dam, such as a dam constructed by the U.S. Army Corps of Engineers or the Bureau of Reclamation
- if it was constructed after 26 August, 1935 and is "located on a non-navigable stream subject to Congress' jurisdiction under the Commerce Clause [or that] affect[s] the interests of interstate or foreign commerce," including any project that feeds power into an interstate power grid

As discussed above, Commission jurisdiction is confined to hydroelectric developments sponsored by nonfederal entities. These include private developers, stockholder-owned utilities, and state or local entities, such as public utility districts. The Commission does not regulate hydroelectric developments sponsored by federal agencies, such as the U.S. Army Corps of Engineers or the Bureau of Reclamation. The Commission does not have authority over river development for purposes other than hydropower.

The issue of Commission jurisdiction is important if you are interested in one of the scores of existing projects around the country that should have a Commission license but does not. As the Commission's jurisdiction has expanded over the years, some dam owners, either deliberately or by oversight, have not come forward and applied for Commission licenses. If you are interested in correcting environmental problems at an unlicensed dam, an appropriate first step might be to encourage the Commission to make it a priority to determine if it has jurisdiction over the project. A subsequent licensing process will provide an opportunity to impose new conditions on the project, to correct environmental problems.

American Rivers and other conservation groups have used this approach with good results as part of an effort to restore Atlantic salmon runs to the Saco River in Maine and New Hampshire. The Swan Falls Dam, near the New Hampshire border, was subject to Commission jurisdiction but had never been licensed. The dam had neither upstream nor downstream fish passage facilities and, in the absence of a Commission licensing proceeding, it would have been difficult to get these facilities installed. The conservation groups wrote to the chairman of the Commission, urging her to make the licensing of Swan Falls Dam a high priority. The Commission granted the request and has now commenced formal proceedings to bring the Swan Falls Dam under license.

FERC'S ORGANIZATION

The Commission is a five-member regulatory body housed, for administrative purposes, within the U.S. Department of Energy. The Commissioners are appointed by the president with the advice and consent of the Senate. The president has responsibility for designating one of the Commissioners as the chairman of the Commission.[5]

In addition to exercising regulatory authority over nonfederal hydroelectric development, the Commission also regulates interstate sales of natural gas and electric power. Indeed, the Commission's other regulatory responsibilities overshadow its hydroelectric work load. This, too, helps to explain the Commission's traditional lack of attention to the environmental consequences of hydroelectric development.

The Commission typically meets once or twice a month at its headquarters in Washington, D.C. While its meetings are generally open, there is rarely an opportunity for members of the public to address the Commission. Most matters that come before the Commission, such as deciding whether or not to issue a license for a hydroelectric project, are acted on without discussion.

Much of the actual work and decision making are carried out by the staff of the Commission. Most of the staff working on hydropower issues are assigned to the Office of Hydropower Licensing. The office has three basic divisions: (1) the Division of Project Review, which reviews project applications; (2) the Division of Project Compliance and Administration, which enforces the terms and conditions of Commission orders and works to bring existing unlicensed projects under Commission jurisdiction; and (3) the Division of Dam Safety and Inspections, which administers the Commission's dam safety program. (See the organizational chart below.) Other Commission staff working on hydropower projects include a number of lawyers, in FERC's Office of General Counsel, who are assigned to hydroelectric issues, and several individuals who serve on the personal staffs of the chairman and the other Commissioners. The Commission also has five regional offices. (See Appendix A for FERC's addresses and telephone numbers.)

FERC ORGANIZATION CHART FOR HYDROPOWER ISSUES

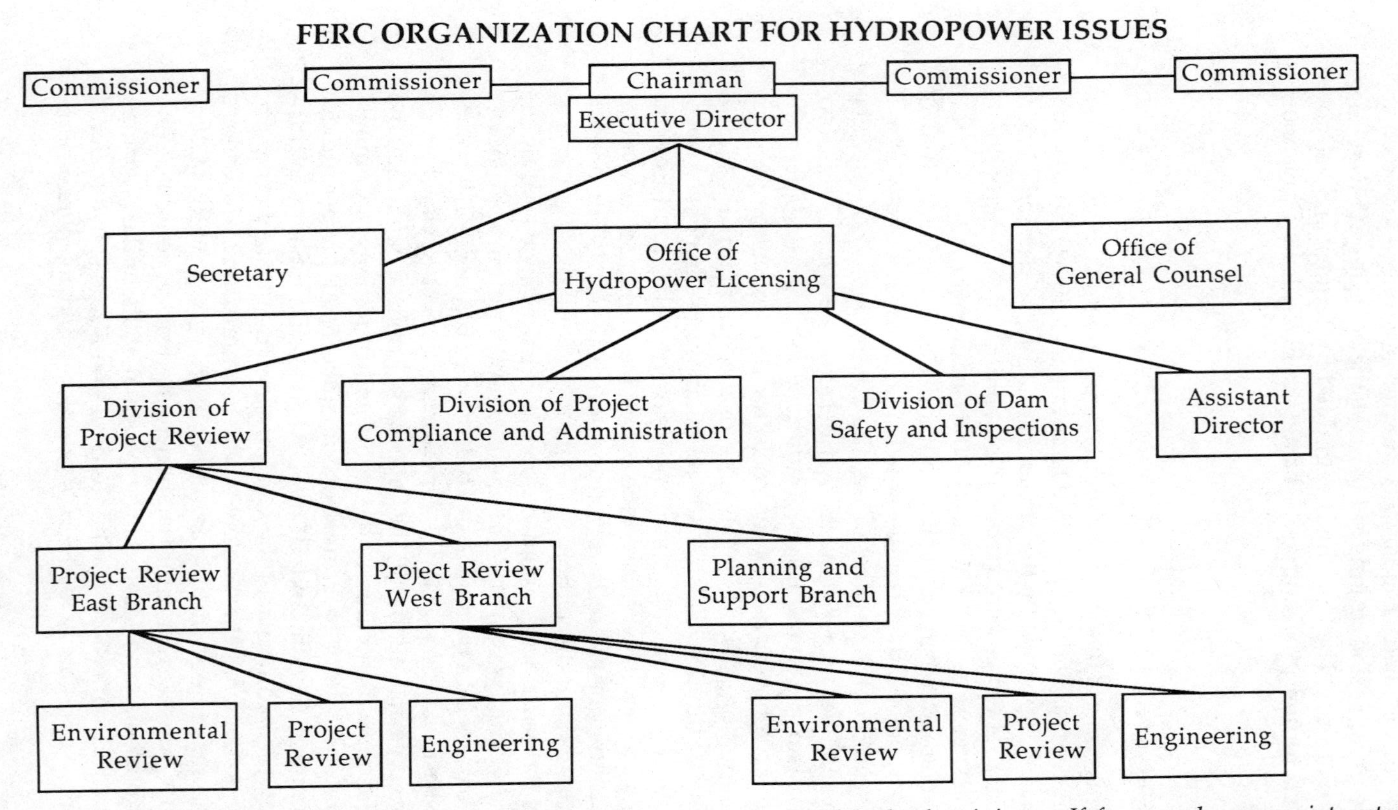

In June 1988, FERC adopted this internal organizational structure to deal with hydroelectric issues. If, for example, you are interested in how a proposed hydroelectric project in New York state might affect a whitewater boating run, call the Office of Hydroelectric Licensing and ask for the director of the East Branch Recreation and Land Use unit. If you are interested in a California project's impact on fisheries, ask for the director of the West Branch Biology unit.

The processing of each proposed hydroelectric project usually is overseen by a project manager. This is the key person to get to know at the Commission. The project manager should be able to tell you the status of the hydroelectric application and to resolve any questions you may have. (See chapter 5 for a discussion of the Commission's *ex parte* rules.)

The Commission has delegated broad decision-making authority with respect to hydroelectric development to the director of the Office of Hydropower Licensing. For example, the director is authorized to act on license, exemption, and preliminary permit applications in every case where there is no formal opposition to the application. (The Commission has, however, reserved the right of applicants and intervenors to appeal the director's decisions to the Commission.) Other examples of delegated authority are laid out in the Commission's regulations.[6]

The Commission also has a permanent cadre of administrative law judges. These judges are available to conduct trial-type proceedings to resolve disputed factual issues. Any party to a hydroelectric proceeding is entitled to an adjudicatory hearing before an administrative law judge if the record contains conflicting evidence on material factual issues and if the party has made a timely request for a hearing.

THE COURTLIKE CHARACTER OF FERC

In dealing with FERC, keep in mind that the Commission views itself as a quasi-judicial, or courtlike, body. Based both on its legislative mandates and on its institutional history, the Commission's central mission is to decide specific cases—whether or not to permit the construction of a proposed project, or what new terms to impose on a project at the time of relicensing. In this respect, FERC is very different from management agencies, such as the Forest Service, or enforcement agencies, such as the Environmental Protection Agency.

Several things follow from the Commission's courtlike character. First, the basic work of the Commission and the staff is to call balls and strikes—to resolve specific cases based on the evidence that applicants, intervenors, and others present. The Commission is not organized to affirmatively seek out potential environmental issues or public concerns not already brought to light. It relies on applicants, federal and state agencies, Indian tribes, and the public to carry the burden of bringing issues forward to the Commission. As a result, the Commission pays little attention to comments submitted by the public not supported by solid factual information and strong legal logic.

Second, very formal rules govern public participation in the Commission's decision-making process. Most federal agencies provide a variety of relatively informal means for citizens to provide oral and written comments on proposed agency action. By contrast, the Commission will not seriously consider the views of any citizen who has not become an official *party*—almost like a litigant in court—to a Commission proceeding. This means that a letter protesting the project is unlikely to have much influence on the Commission's decision. Although the Commission will accept such a letter of comment, you will be a party to the proceeding only if you file a motion to intervene.

Furthermore, like a court, the Commission enforces strict procedural rules. For example, if a citizen fails to file a *motion to intervene* by the official deadline (see chapter 5), he or she must file a separate *motion for leave to file a late intervention motion,* which the Commission might well deny. The Commission has been known to reject motions to intervene when the moving citizens failed to include a *certificate of service* or adequately describe their interest in a proceeding.

On the other hand, the Commission has granted some motions for late intervention when there were only weak justifications for not having met the original filing deadline. The lesson is that the Commission acts on each petition on a case-by-case basis, and it may well be worth filing even if you are late.

Susan Rand

Bulls Bridge Dam on the Housatonic River in Connecticut. A proposed project expansion threatened the white-water boating run below the dam.

THE STORY OF BULLS BRIDGE GORGE OR IT'S NOT NECESSARILY TOO LATE

The successful efforts of paddlers to protect white-water boating in Bulls Bridge gorge demonstrate the value of effective, even if tardy, intervention at FERC.

Bulls Bridge Gorge is located on the Housatonic River in Connecticut. It is a beautiful stretch of exciting class three to class four white water, a rarity in southwestern New England. Connecticut Light and Power filed an application with FERC to add generating capacity to the existing Bulls Bridge project above the gorge. The company proposed to more than double the volume of the flow in the bypass canal, which would have reduced the number of boatable days on the river from approximately 90 to approximately 30.

For a number of years, local paddlers, led by the Housatonic Area Canoe and Kayak Squad (HACKS), had been attempting to negotiate the issue of white-water releases with the company. Frustrated by the direction of these negotiations, the local paddlers enlisted the aid of the American Whitewater Affiliation and a *pro bono* attorney experienced in hydro matters to represent them before FERC.

The problem was that the company had filed the application two and a half years before the paddlers had enlisted expert assistance, and the formal deadline for filing motions to intervene had come and gone. Nonetheless, the paddlers filed a motion for late intervention (see chapter 5) and raised a host of issues concerning white-water recreation that had been completely ignored in the FERC process until then. The company objected that the paddlers had no good cause for late intervention and that state and federal resource agencies could adequately protect the paddlers' interests in the FERC process.

Surprisingly, given the Commission's strict rules on late filings, FERC issued an order allowing the American Whitewater Affiliation, HACKS, and the American Canoe Association to participate as formal parties to the proceeding.

Connecticut Light and Power was already embroiled in a difficult struggle with state and federal environmental authorities over how to dispose of dangerous polychlorinated biphenyls (PCBs) that had been found in the power canal. Faced also with this strong opposition from HACKS and its allies, the power company announced that it was abandoning the proposed expansion. The company's official position was that the economic benefits of the project had deteriorated since the time of the original studies on the project. It is a good bet, however, that the paddlers' successful intervention motivated that decision.

A WORD ABOUT PURPA

The Public Utility Regulatory Policies Act (PURPA), enacted in 1978, has been the single most important force driving the hydropower gold rush of the 1980s.[7] PURPA created the opportunity for independent power producers—as distinguished from public and private utilities obligated to serve retail power customers—to enter the business of developing hydroelectric projects (as well as certain other types of energy projects). The details of how PURPA works are beyond the scope of this handbook, but a thumbnail sketch of PURPA will be useful in understanding the regulatory forces driving hydroelectric development. Also, as a result of the enactment of ECPA, some unique environmental requirements now apply to projects developed under PURPA (see chapter 3).

Congress enacted PURPA in 1978 in response to prevalent concerns about the adequacy and security of the nation's energy supplies. In order to reduce the country's dependence on fossil fuel, and on imported oil in particular, Congress sought to encourage the construction of a variety of "small" power facilities by independent power producers. (Congress defined *small* power facilities as those having a capacity of less than 80 megawatts, which in the case of hydro projects is actually quite large indeed.) To achieve this objective, Congress provided various incentives for independent power producers, including the guaranteed right to sell to utilities the power they produce.

More specifically, PURPA directed the Commission to promulgate rules requiring public utilities to purchase power from so-called qualifying facilities at defined rates. These rates were not to exceed the "full avoided cost," or what it would have cost the utilities to produce the electrical energy themselves. PURPA also gave the Commission the authority to require electric utilities to allow qualifying facilities to "interconnect" with the utilities' transmission system. In addition, PURPA allowed the Commission to exempt qualifying facilities from the type of extensive state and federal regulation that applies to utilities.

Congress clearly intended so-called PURPA benefits to be available to developers of hydropower at existing dams, for example, abandoned industrial dams with unexploited hydropower potential. On the other hand, Congress indicated that it did not intend for PURPA benefits to apply to hydroelectric development of pristine, free-flowing rivers. While the Commission's original draft of the PURPA rules adhered to Congress's intent, the final version of the rules, adopted in 1980, extended PURPA benefits to all types of hydroelectric development.[8] The Commission's broad interpretation of PURPA was instrumental in spurring hundreds of hydropower proposals involving new dams and diversions (structures which do not block the entire watercourse but divert water out of a river into a canal or penstock to generate hydropower). Interest for such development focused primarily on California, the Pacific Northwest, and the Northeast.

As discussed in more detail in the following chapters, Congress amended PURPA, in the Electric Consumers Protection Act of 1986, to clarify its original intent to restrict the availability of PURPA benefits to projects involving existing dams and diversions. Congress established a complete moratorium (subject to various grandfather provisions) on PURPA benefits for new dam and diversion projects, and it ordered the Commission to complete a study of the entire question. At present, the moratorium is scheduled to expire at the end of 1989, by which time Congress should have had an opportunity to consider the issue. If Congress chooses not to act by the end of 1989, PURPA benefits will again become available for new dam and diversion projects, but only if the projects meet stringent new environmental requirements.

REFERENCES

1. See 41 Stat. 1063. The Federal Power Act appears in the United States Code at 18 U.S.C. § 791 et seq.
2. 49 Stat. 863.
3. Pub. L. No. 99–495, 100 Stat. 1244.
4. 16 U.S.C. § 797(e).
5. 16 U.S.C. § 792.
6. See 18 C.F.R. § 375.314.
7. Pub. L. No. 95–617. The PURPA provisions relating to hydropower appear at 16 U.S.C. §§ 824–824a-4, 824d, 824i–824k, 825d, 2701–2707.
8. The Commission's regulations implementing sections 201 and 210 of PURPA appear in 18 C.F.R. § 292.101 et seq.

Note: see Appendix H for an explanation of how to find and use legal documents upon which FERC relies.

Chapter 2

An Overview of the Regulatory Process

FERC's preliminary permits, licenses, and exemptions • Relicensing options for existing projects • FERC's decision process

YOUR ABILITY TO INFLUENCE the Commission's decision making on a project application depends on your understanding of how the Commission works. This chapter describes the types of applications required by the Commission, lists the options in the relicensing of existing projects, and explains how the Commission staff processes hydropower applications.

THE STEPS IN ORIGINAL LICENSING

Preliminary Permits

A developer interested in a potential site for a hydroelectric project can first apply to the Commission for a *preliminary permit.*[1] A preliminary permit grants the developer the exclusive right to file a license application for the site during the term of that permit. The purpose of a preliminary permit is to protect a developer, who is investing time and money to study a potential site, from the risk that another developer will file a competing license application for the same site. A preliminary permit does *not* authorize construction of a project; it merely authorizes the permittee to study a site and develop a proposal.

A developer is not actually required to obtain a preliminary permit before filing an application for a license. In practice, however, most developers get preliminary permits before putting in the money, time, and effort required to develop a license application.

The preparations required for a preliminary permit are fairly modest, and Commission approval is almost automatic. The basic requirements are a description of the proposed project and a description of the environmental, engineering, financial, and other studies the applicant intends to conduct to determine whether or not to proceed with the project.

Sometimes the Commission is faced with competing preliminary permit applications for the same site. The Commission resolves competitions between two permit applicants using the following order of priority: (1) a state or municipal utility, (2) the applicant who would "best serve the public interest," and (3) the applicant who filed first in time.[2] In some instances a developer will file a license application for a site for which a preliminary permit application has been filed but has not yet been approved. In such a case the Commission will dismiss the preliminary permit request in favor of the license application unless the preliminary permit applicant comes forward with information to "substantiate" the technical, environmental, and economic aspects of the project.[3]

The Commission usually issues preliminary permits for a period of three years. Since the Federal Power Act states that a preliminary permit may be issued for a period or periods "not exceeding a total of three years," a developer who receives a three-year permit is barred from obtaining an extension of the original term.[4] The Commission has ruled, however, that a permittee can apply for a second preliminary permit, provided that the permittee is proceeding with the project proposal diligently and in good faith, that there is a new opportunity for public comment, and that other potential developers can apply for the new permit.[5]

Preliminary permits typically require the permit holder to file reports with the Commission every six months. The reports must describe the progress of the permittee's studies on the environmental, engineering, and financial feasibility of the proposed project.

Licenses

Once a potential site has been evaluated, the developer can apply to the Commission for a *project license.*[6] Assuming the project is subject to Commission jurisdiction (see chapter 1), a Commission license or an exemption from the licensing requirements (see Exemptions on page 31) is mandatory for lawful construction and operation of the project. In addition to authorizing the construction and operation of the project, a license contains detailed conditions and terms. These include, for example, restrictions designed to protect environmental and recreational resources during the project's operation. Many standard terms and conditions are used in all FERC licenses. Other terms and conditions are tailor made to the specific project and river.[7] (See Appendix D for an example of a hydroelectric project license.)

When two or more applicants compete for an original license to construct a project, a public utility (for example, a city-owned utility) is entitled to preference over a private entity if the applications are otherwise equal.[8] (This so-called municipal preference applies only in original licensing proceedings, not in relicensing proceedings for existing projects.) If the municipal preference does not apply, the Commission is required to select the proposal that best meets the licensing standards under the Federal Power Act.

The volume and detail of information that must be submitted to support a license application are far greater than what is required for a preliminary permit. In general, the license application is supposed to provide the Commission sufficient information to determine whether the project is feasible from the economic and engineering standpoints. It should also contain an explanation of the expected environmental consequences of the project and a description of the proposed conditions that will be applied to prevent or minimize adverse environmental effects.

The Commission's regulations divide license applications into several categories. These include the following:

- "major unconstructed" projects, that is, projects of more than 1.5-megawatt capacity that would utilize newly constructed dams[9]
- "major" projects at existing dams, that is, projects of more than 1.5-megawatt capacity that would exploit the water power potential of existing dams[10]
- "minor" projects, that is, projects of 1.5-megawatt capacity or less that would involve either the construction of new dams or the installation of hydroelectric facilities at existing dams[11]

The Commission's license application requirements are more detailed for major projects than for minor ones.[12] (As discussed later, these categories are also important in understanding other parts of the Commission's regulations.)

Licenses are issued for a fixed term, up to a maximum of 50 years.[13] A license is like a contract, entitling the licensee to own and operate the project in accordance with the license conditions for the term of the license. Unless the Commission expressly has reserved the power to amend the license, it ordinarily cannot be altered without the agreement of the licensee. In practice, though, the Commission can unilaterally amend many outstanding licenses because the licenses include *reopener* clauses. These expressly authorize the Commission to modify the licenses as conditions warrant.[14]

The Federal Power Act states that a developer must commence construction by the date set in the license (which cannot be more than two years from the date of issuance) and that the developer thereafter must proceed with construction "in good faith and with due diligence."[15] The act also states that the deadline for commencing construction may be extended only once, for a maximum of two additional years. If the licensee fails to commence construction within the time prescribed, the Commission can rescind the license.

Exemptions

In some cases, a developer may seek authorization to construct and operate a project by applying for an *exemption* from the Federal Power Act's licensing requirements.[16] An exemption literally excuses the applicant from the requirement of obtaining a license from the Commission. However, a developer seeking an exemption must satisfy requirements that are analogous to, though less rigorous than, those for a license application.[17]

A developer may seek an exemption from the Federal Power Act's licensing requirements for the following types of projects:

- a project utilizing an existing conduit (e.g., irrigation or water supply canal) on nonfederal lands, so long as the conduit is operated primarily for agricultural, municipal, or industrial water supply purposes and the project has a capacity of 15 megawatts or less (or 40 megawatts or less in the case of a facility operated by a state or local government solely for municipal water supply purposes)[18]
- a project with a generating capacity of not more than 5 megawatts that either is located at an existing dam or utilizes "natural water features for the generation of electricity, without the need for any dam or impoundment"[19]

The Commission's implementation of the "natural water features" exemption has been controversial. Congress adopted this provision to encourage the construction of projects that exploit the natural hydropower potential of elevated lakes or waterfalls. In 1982, however, the Commission attempted to enlarge the number of projects eligible for the exemption, by adopting a rule that defined *natural water features* as including "diversions structures" that are under ten feet in height and that impound no more than two acre-feet of water. In *Tulalip Tribes v. FERC* a federal court invalidated the FERC rule on the ground that it was inconsistent with the language of the Public Utility Regulatory Policies Act (PURPA).[20] The court held that Congress had clearly intended to exclude from the exemption process any project that uses any dam or impoundment.

An important condition attached to every exemption is that the developer must agree to construct the project in accordance with terms and conditions set by fish and wildlife agencies. Specifically, the developer must agree to abide by conditions designed by the U.S. Fish and Wildlife Service, the National Marine Fisheries Service, or state fish and wildlife agencies "to prevent loss of, or damage to" fish and wildlife resources.[21]

RELICENSING

The Federal Power Act provides that the maximum term of a hydroelectric license is 50 years. During the deliberations in 1920 over the Federal Water Power Act, there was sharp debate between those who wished to encourage maximum private development of rivers and those who wished to retain public control. The 50-year term for hydroelectric licenses represents a compromise between these two positions. It guarantees that a developer will be able to operate a project for a sufficient number of years to recoup the original investment and make a reasonable profit. On the other hand, the limited license term ensures that licenses will automatically terminate and that the Commission will be forced periodically to review how a river should be used to best serve the public interest.

The formal review of an existing project at the expiration of its

John D. Echeverria

The Cataract Project on the Saco River in Maine. Relicensing provides an opportunity to improve existing dams by installing fishways, like the one in this photograph.

original license term is known as *relicensing*. But there are actually several different regulatory options available to the Commission.

First, the Commission can relicense the project, that is, issue either a new license to the original licensee or a new license to a new owner, authorizing the continued operation of the project to generate hydropower.[22] The Commission has a duty, especially in light of the Electric Consumers Protection Act, to ensure that older projects are modified to achieve a better balance between power generation and protection of environmental resources. Second, the Commission or another federal agency can recommend a *takeover.* That is, they can suggest to Congress that the project be taken over from the licensee and operated by the government.[23] Finally, a private entity or local government can apply to the Commission for a *nonpower license* for a project, to facilitate the conversion of the project to some other use. The project's reservoir could be managed exclusively for recreational use, for example. Possibly, the dam could be removed entirely in order to re-create a free-flowing river.[24]

So far as we are aware, federal takeover and nonpower licensing have been attempted on only a few occasions,[25] and they have never succeeded to date. However, conservation groups in Maine are currently exploring the possibility of seeking a nonpower license for the purpose of removing a dam that blocks anadromous fish runs on the Kennebec River. Also, the Michigan Department of Natural Resources has stated publicly that it is considering seeking a nonpower license for certain existing dams in that state.

Five years before the original license expires, a project owner must file a notice with the Commission stating whether the owner intends to seek a new license for the project. The deadline for submitting an actual relicense application—both for the incumbent and for a potential new owner—is two years before the expiration of the original license. The procedures and standards governing the relicensing of existing projects are discussed in detail in chapter 4.

HOW THE APPLICATION REVIEW PROCESS WORKS

When it receives an application for a preliminary permit, license, or exemption, the Commission assigns it a four-digit or five-digit *project number.* This number is used to keep track of the application as it wends its way through the regulatory process. At the same time, the application is placed on the shelves of the FERC public information room, where it is available for public inspection (usually for several months). (See Appendix A for the address of FERC's public information room.)

Acceptance for Filing

The staff reviews the application to determine whether it contains sufficient information to satisfy the Commission's application requirements. Once satisfied that the application meets the requirements, the staff *accepts the application for filing*. If the application is deficient in any respect, the staff writes to the applicant and requests that the deficiencies be corrected. The Commission sometimes issues more than one deficiency letter on a particular application and usually requests additional information from an applicant in order to complete the processing of the application. In extreme cases, the Commission may simply reject an application. All told, the process of reviewing the adequacy of the application and obtaining the additional information can take several months or, in rare cases, several years.

In order for FERC to accept it as properly filed, an application must include certain information specified by the Commission's regulations. The following table summarizes that information, by type of application, and provides cross-references to the Code of Federal Regulations, volume 18. For example, in an application for a major new project (greater than five megawatts of capacity, not at an existing dam), the project's design and operation are addressed in exhibit A (physical description of the project site and facilities), exhibit B (statement of alternative designs that were considered, estimation of energy production and water flow), exhibit F (design drawings), and exhibit G (map). The information which must be included in these exhibits is specified in 18 C.F.R. §§ 4.41b, c, g, and h.

Notice in the *Federal Register* and Newspapers

A notice describing the project is then published in the *Federal Register*, a daily publication of regulations and legal notices issued by federal agencies. At the same time, the staff arranges for the notice to be published in local newspapers in the locale of the proposed project. The notice includes a *comment date*, which is the deadline for filing a *motion to intervene* in the application proceeding. As discussed in detail in chapter 5, a member of the public must file a motion to intervene in order to become an official participant in the proceeding; as a practical matter, intervenor status is essential in order to have real influence over the outcome of the process.

The Commission grants almost all preliminary permit applications. FERC generally ignores environmentally based objections to the issu-

Information Required in FERC Applications
(Exhibit Numbers or Letters, with Code of Federal Regulations References in Parentheses)

	Project Design and Operation	*Construction Plan*	*Project Costs and Finance*	*Environmental Impacts*
Preliminary Permit	1 (§ 4.81(b)) 4 (§ 4.81(e))	2 (field studies only) (§ 4.81(c))	3 (§ 4.81(d))	2 (field studies only) (§ 4.81(c))
License				
Major new project or major modification, greater than 5 MW	A (§ 4.41(b)) B (§ 4.41(c)) F (§ 4.41(g)) G (§ 4.41(h))	C (§ 4.41(d))	D (§ 4.41(e))	E (§ 4.41(f))
Major project at existing dam, greater than 5 MW	A (§ 4.51(b)) B (§ 4.51(c)) F (§ 4.51(g)) G (§ 4.51(h))	C (§ 4.51(d))	D (§ 4.51(e))	E (§ 4.51(f))
Minor project, or major project less than 5 MW	A (§ 4.61(c)) F (§ 4.61(e)) G (§ 4.61(f))	A (§§ 4.61(c)(1)(i)(viii))	A (§§ 4.61(c)(1)(ix)(c)(2))	E (§ 4.61(d))
Amendment	Exhibits depend on scope of project, track above requirements. See 18 C.F.R. § 4.201.			
Relicensing	The application for a new license for an existing project will fall into one of the three categories for initial licenses and, for a major project, include a special Exhibit H (§ 16.10).			
Exemption				
Small conduit project	A (§ 4.92(c)) B (§ 4.92(d)) G (§ 4.92(f))	A (§§ 4.92(c)(8)(11))		E (§ 4.92(e))
Case-specific exemption	A (§ 4.107(c)) B (§ 4.107(d)) G (§ 4.107(f))	A (§§ 4.107(c)(7)(8))		E (§ 4.107(e))

ance of preliminary permits. The reasoning of the Commission is that environmental objections are irrelevant at that stage because the permit authorizes the applicant only to study the project, not to commence construction.

Consultation with Other Agencies

In 1985, the Commission adopted its "pre-filing consultation requirements" for license applications.[26] These regulations require an applicant to consult closely with state and federal resource agencies before filing an application for a license or for an exemption. The consultation process is intended to ensure that resource issues are identified early in the review process, that the applicant conducts necessary studies, and that any conflicts between the applicant and the resource agencies are resolved to the extent possible, before the Commission acts on the application. Unfortunately, state and federal resource agencies already are overburdened with work, and all relevant agencies may not be able to participate effectively at this stage. Also, the public is largely excluded from this process.

In the first stage of the consultation process, the applicant must contact appropriate resource agencies and provide each of them detailed information about the project. Appropriate resource agencies include, for example, the state agency that issues water quality certifications under section 401 of the Clean Water Act (see chapter 7). Also included are the Environmental Protection Agency, state and federal fish and wildlife agencies, recreation agencies, the state historic preservation office, and the Bureau of Land Management or the U.S. Forest Service if the project is located on lands administered by either of these agencies. No representative of FERC directly participates in the process. The purpose of the first stage of the consultation process is not only to educate the agencies about the proposed project, but also to obtain their views on what environmental studies need to be completed before an acceptable license application can be prepared.

In the second stage, the applicant responds to all reasonable study requests by the agencies and conducts any other studies that are "necessary for the Commission to make an informed decision regarding the merits of the application."[27] When the studies are complete, the applicant provides copies of the study reports and a copy of a draft license application to each of the resource agencies, along with a written request to each agency for review and comment. The agencies have between 30 and 60 days to comment.[28]

In the third stage of consultation, the applicant serves on each of the consulted agencies a copy of the final license application. An *exhibit* to the application must explain how the applicant complied with the consultation requirements, and it must include copies of correspondence from the agencies containing comments, recommendations, and proposed conditions.

Even though it precedes formal Commission review of a license application, the prefiling consultation is a crucial part of the decision-making process. Commonly, the applicant and the agencies use the consultation process as an opportunity to reach substantive agreements on how to deal with a variety of critical resource issues. This arrangement presents a difficult problem for members of the public, since the Commission's consultation regulations do not provide for public participation at this stage. You have several legal options. One is to ignore the consultation process and then intervene once the completed application has been filed with the Commission. The danger is that, if you or other members of the public have not been a party to the years of negotiations leading up to the filing of the application, your concerns may be overlooked or simply dismissed as inconsistent with the agreements already negotiated.

Another option, and a potential solution to the problem, is to inform the resource agencies about public concerns so that the agencies can raise those concerns in the consultation process. The weakness of this approach is that the resource agencies may, for any of a host of possible reasons, not agree with the public or fail to present an issue as forcefully as you would. A third potential option is to simply tell the agencies and/or the applicant that you wish to be a party to the consultation process. FERC's regulations do not prohibit public participation in the consultation process.

The Commission's Environmental Analysis

Once a license application is accepted for filing by the Commission, the Commission must make some important decisions about how the application will be processed. The first issue is the level of environmental analysis the Commission needs to perform under the National Environmental Policy Act.[29] If the application is not controversial and appears to present relatively modest environmental risks, the Commission staff is likely to prepare an environmental assessment (EA) of the project. An EA is generally a short document (approximately ten pages) that analyzes, in summary fashion, the probable environmental effects of and alternatives to the proposed action. If the Commission relies on an EA, there is generally no opportunity for the public to comment on the

HYDROELECTRIC LICENSING PROCESS

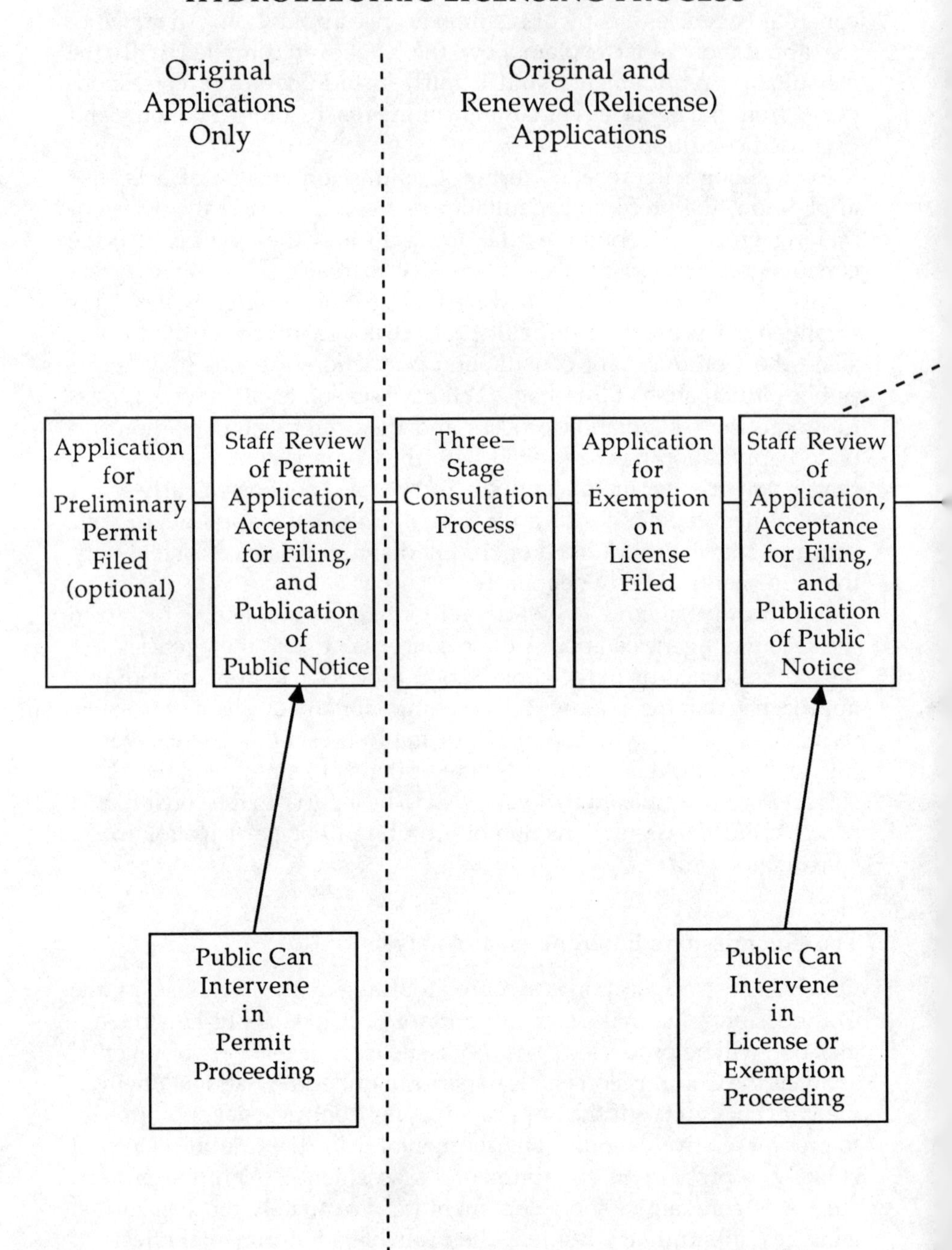

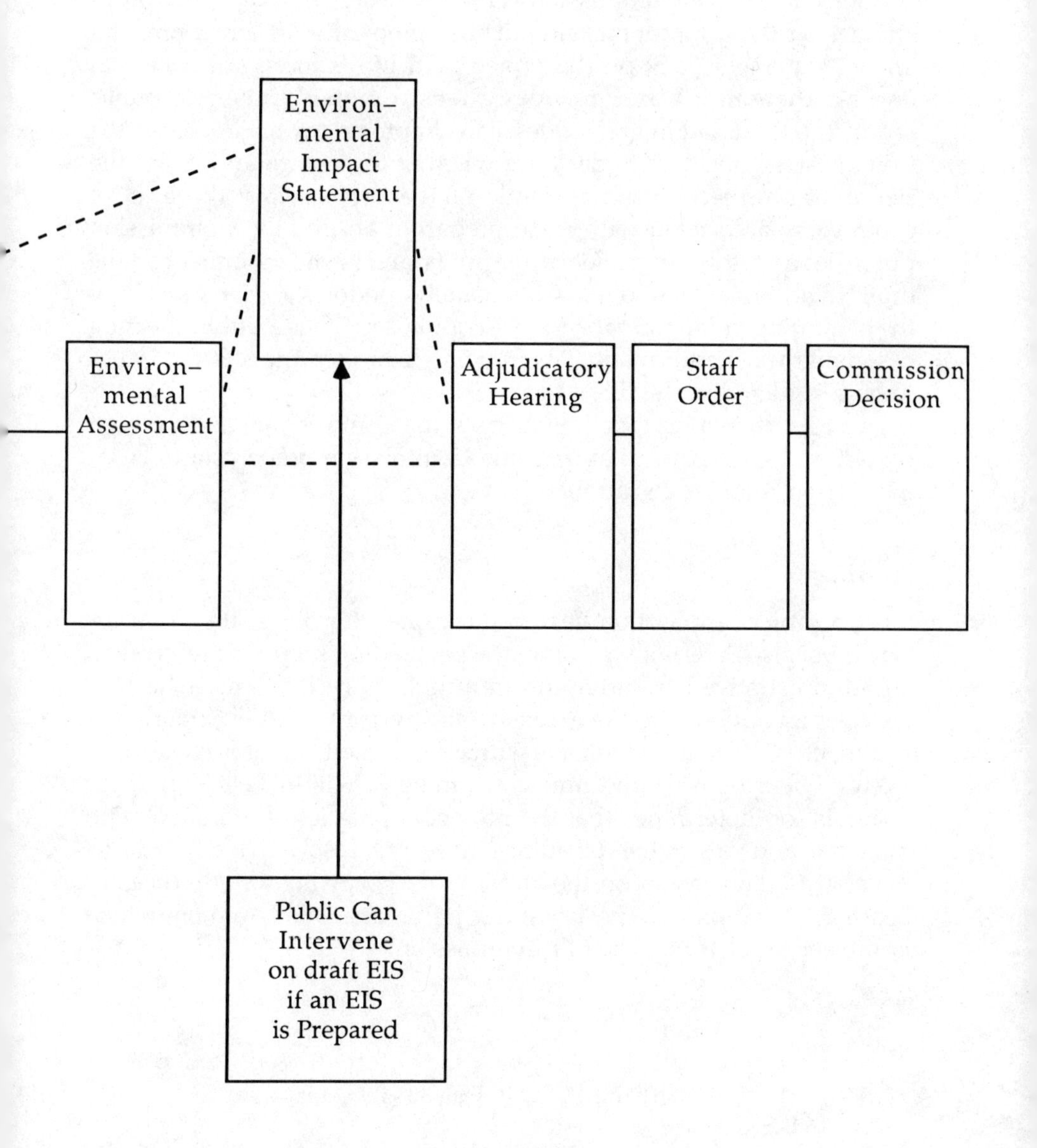
Environ–
mental
Impact
Statement
Environ–
mental
Assessment
Adjudicatory
Hearing
Staff
Order
Commission
Decision
Public Can
Intervene
on draft EIS
if an EIS
is Prepared

environmental analysis before the Commission reaches a decision on the application.

Alternatively, the Commission staff may decide that it cannot properly act on the application without first preparing an environmental impact statement (EIS) on the project. An EIS is more elaborate and detailed than an EA and provides extensive opportunities for public input. As discussed in greater detail in chapter 6, the major factors the Commission considers in deciding whether to prepare an EIS are the size of the project, the magnitude of the project, and the level of controversy associated with it. In preparing an EIS, the Commission publishes a draft version and invites public and agency comment on the draft. In addition, the Commission usually conducts a *scoping session* at the beginning of the EIS process. A scoping session is a public meeting designed to elicit information about possible project impacts and issues to be considered in the EIS.

Due to stringent budget limitations at the Commission, and the high cost of an EIS relative to an EA, the Commission prepares an EA far more often than it does an EIS.

Hearing

Once a project application has been accepted for filing, the Commission must also determine whether the proceeding should be referred to an administrative law judge for an adjudicatory hearing. A hearing provides an opportunity for extensive *discovery* into the developer's files and for cross-examination of live witnesses. An adjudicatory hearing is likely to be expensive and time-consuming; it will be held only if the Commission determines that the project application, the intervention motions, and any other pleadings have not resolved issues of fact material to its decision on the application. Following an adjudicatory hearing, the administrative law judge issues a detailed opinion, which is subject to review by the full Commission.

REFERENCES

1. See 16 U.S.C. § 797(f) and 18 C.F.R. §§ 4.30–4.39, 4.80–4.86.
2. See 16 U.S.C. § 800(a).
3. See *In re Consumnes River Water and Power Authority,* 42 FERC ¶ 61,069 (CCH) (25 Jan. 1988).
4. 16 U.S.C. § 798.
5. See *In re City of Redding,* 33 FERC ¶ 61,109 (CCH) (15 Oct. 1985).
6. See 16 U.S.C. § 797(e) and 18 C.F.R. §§ 4.30–4.61.

7. Citations to the Commission's standard license and permit conditions are listed in 18 C.F.R. § 2.9.
8. 16 U.S.C. § 800(a).
9. 18 C.F.R. § 4.30(b)(15).
10. 18 C.F.R. § 4.30(b)(16).
11. 18 C.F.R. § 4.30(b)(17).
12. See 18 C.F.R. § 4.41 (major unconstructed project), 18 C.F.R. § 4.51 (major project—existing dam), 18 C.F.R. § 4.61 (minor project and major project five megawatts or less).
13. See 18 U.S.C. § 799.
14. See *In re Central Nebraska Public Power and Irrigation District,* 43 FERC ¶ 61,225 (CCH) [5 May 1988 (discussing Commission requirement of "reservation of authority" in order to unilaterally amend license)].
15. 16 U.S.C. § 806.
16. See 18 C.F.R. §§ 4.90–4.113.
17. 18 C.F.R. § 4.92 (conduit facilities), 18 C.F.R. §§ 4.107–4.108 (projects of five megawatts or less).
18. See 16 U.S.C. § 823a and 18 C.F.R. §§ 4.90–4.96.
19. See 16 U.S.C. §§ 2701–2708 and 18 C.F.R. §§ 4.101–4.113.
20. 732 F.2d 1451 (9th Cir. 1984).
21. 16 U.S.C. § 823a(c).
22. See 16 U.S.C. § 808.
23. See 16 U.S.C. § 807.
24. See 16 U.S.C. § 808(f).
25. See *Lac Court Oreilles Band of Lake Superior Chippewa Indians v. FPC,* 510 F.2d 198 (D.C. Cir. 1975) (federal takeover proposal) and *In re Escondido Mutual Water Company,* 6 FERC ¶ 61,189 (CCH) (20 Feb.1979) (nonpower license application and federal takeover proposal).
26. See 18 C.F.R. § 4.38.
27. 18 C.F.R. § 4.38(b)(2).
28. 18 C.F.R. § 4.38(b)(2)(iv).
29. The Commission's regulations implementing the National Environmental Policy Act are at 18 C.F.R. §§ 380.1–380.11.

Chapter 3

New Projects: Standards and Procedures

FERC's application of the public interest standard, the equal consideration test, and the comprehensive planning provision • FERC's consultation with fish and wildlife agencies, federal land management agencies, and others • FERC's environmental analysis • Environmental restrictions on PURPA projects

THE COMMISSION IS REQUIRED by law to adhere to specific standards and procedures in deciding whether and under what conditions to permit the construction and operation of a hydroelectric project. Some of these standards and procedures apply to all hydroelectric projects; others apply only to particular kinds of projects. Another set of requirements determines whether projects are entitled to the benefits of the Public Utility Regulatory Policies Act (PURPA). Understanding these rules of the licensing process is essential to identifying the issues that are—and are not—worth raising before the Commission.

This chapter discusses the standards and procedures that govern an original hydroelectric project application to the Commission. This includes new construction on a free-flowing river as well as the installation of hydroelectric equipment at an existing or refurbished dam. The standards and procedures for relicensing—when a developer seeks a license for a project whose original license is expiring—are discussed in the following chapter.

A word about organization: this chapter is organized not by issue (e.g., a project's environmental impacts), but by statutory provision. Each such provision discussed here creates a standard that affects FERC's substantive review of an application. You will notice that many provisions overlap: for example, the *comprehensive planning* standard, the *equal consideration* standard, and the National Environmental Policy Act all oblige FERC to consider, in various ways, the environmental impacts of a proposed project. This chapter does not present such related provisions as a group, however, so that you can better understand each in isolation and get a sense of the unique features of each. Chapter 6, which discusses issues you might raise in an intervention motion, groups statutory provisions according to the issues they govern.

FERC'S LICENSING STANDARDS

The "Public Interest" and the New "Equal Consideration" Tests

The basic standard for deciding whether or not to allow hydroelectric development is set forth in section 4(e) of the Federal Power Act.[1] To oversimplify a bit, section 4(e) directs the Commission to review an application based on what it determines is "desirable and justified in the public interest."

The United States Supreme Court, in *Udall v. FPC*[2] emphasized the breadth of this standard. The court stated that

> the test is whether the project will be in the public interest. And the determination can be made only after an exploration of all issues relevant to the public interest, including future power demand and supply, alternate sources of power, the public interest in preserving reaches of wild rivers and wilderness areas, the preservation of anadromous fish for commercial and recreational purposes, and the protection of wildlife.[3]

This language, from the pen of the environment's greatest friend on the Supreme Court, William O. Douglas, was meant to focus the Commission's attention on the public benefits of free-flowing rivers. Unfortunately, the broad scope of the "public interest" test has worked against river preservation. The Commission has rejected on environmental grounds only a small handful of projects, while it has authorized hundreds of projects in the last few years alone. In attempting to define what best serves the public interest, the Commission traditionally has given far less attention to river protection and preservation than it has to river development.

In 1986, Congress strengthened the language of section 4(e) with the enactment of the Electric Consumers Protection Act of 1986 (ECPA). The amendment was intended to make clear that, as Justice Douglas suggested, the public interest will sometimes be served best by preserving free-flowing rivers rather than developing them for hydropower. ECPA adds a new sentence to section 4(e) that states:

> in deciding whether to issue any license [for a hydroelectric project], the Commission, in addition to the power and development purposes for which licenses are issued, shall give *equal consideration* to the purposes of energy conservation, the protection, mitigation of damage to, and enhancement of, fish and wildlife (including related spawning grounds and habitat), the protection of recreational opportunities, and the preservation of other aspects of environmental quality. (Emphasis added.)

This new "equal consideration" test is probably the single most important standard for river conservationists to keep in mind in dealing with the Commission. Basically, it means that all river uses are supposed to have the same weight in the decision-making process. Congress did not, however, go so far as to require equal treatment of all river uses in each individual licensing proceeding, as some advocated during the deliberations on ECPA.

Aside from establishing the general proposition that conservation values deserve equal consideration with power development, the new language in section 4(e) identifies several specific items that must be considered. These include the following:

- *Energy conservation.* If an investment in a proposed hydroelectric project could be used, instead, to implement conservation measures that would provide the same or more power, ECPA strongly supports the argument that the Commission should deny the application. Feasible energy conservation measures are likely to include such things as more efficient appliances and better insulation. When energy conservation is competitive with hydroelectric development on an economic basis, preserving river resources will almost certainly serve the public interest better than development.
- *Protection, mitigation of damage to, and enhancement of fish and wildlife (including related spawning grounds and habitat).* Protection of fish and wildlife—and their habitat—is now an express goal of the Federal Power Act. Rivers and adjacent lands constitute an unusually productive and diverse ecological zone, and the Commission has a duty to protect—and enhance—these resources.
- *Protection of recreational opportunities.* Rivers and their landscapes provide the environment with a wide variety of recreational experiences, such as fishing, boating, bird-watching, picnicking, and hiking. All of these public uses of river resources now have the same legal protection as development opportunities. When the Commission approves a project, it has a duty to consider how to mitigate damage to the river, and even improve it, to maintain its recreational value.
- *Other aspects of environmental quality.* This catchall provision encompasses every conceivable aspect of environmental quality—water quality, aesthetics, historic resources, noise pollution, and so on. If your environmental concern about a project does not fit into any neat category, this provision guarantees you the right to raise it and obligates the Commission to consider it.

NO WAY BIG A

Some hydro battles are bigger than others. One of the very biggest of the 1980s was the successful effort by a broad coalition of Maine conservation groups. With support from national conservation groups led by American Rivers, the coalition defeated the proposal of the Great Northern Paper Company to build the 40-megawatt Big A hydroelectric project on the West Branch of the Penobscot River in Maine.

The company argued that the dam was necessary to supply inexpensive electricity for its mills, and thus to protect the jobs that depended on the mills. Conservationists replied that electricity could be supplied more cheaply, that building the dam would not protect jobs, and that the project would destroy four miles of the finest landlocked salmon fishery in the Northeast and one of the East's premier white-water rafting and boating runs. Bumper stickers on thousands of automobiles across the state demonstrated the depth of feeling about the issue: "All the Way Big A" versus "No Way Big A."

Great Northern sought regulatory approval from three different agencies: (1) the Maine Land Use Regulatory Commission, because the project was located in an unincorporated area of Maine; (2) the Maine Board of Environmental Protection, which exercised state authority over the project under section 401 of the Clean Water Act; and (3) FERC. The company elected to try to complete the state regulatory process first; they went initially to the Land Use Regulatory Commission (LURC). Although the company argued that LURC approval was not strictly required (see chapter 7 for a discussion of the federal preemption issue), the company elected to cooperate in the state regulatory process.

There were more than seven weeks of administrative hearings before LURC. Staff lawyers for the Natural Resources Council of Maine and the Maine Audubon Society, the leaders of the effort, represented the state coalition in all the proceedings. They were assisted by lawyers from the Conservation Law Foundation in Boston and from the prestigious Washington, D.C., law firm of Wilmer Cutler and Pickering, which volunteered its legal services on a *pro bono* basis. The coalition presented extensive testimony on the value of the landlocked salmon fishery and the relative significance of the threatened white-water boating opportunities. They also pointed out economical alternatives that the company could pursue to generate electricity at lesser environmental cost. In the end, LURC approved the project, subject to a number of stringent conditions, by a vote of four to three. Big A was alive, but limping.

The action then shifted to the Maine Board of Environmental Protection, the state agency responsible for deciding whether or not to grant certification for federally licensed projects. (See chapter 7 for a discussion of section 401 of the Clean Water Act.) After extensive briefing and a hearing on the issues, the board denied the certification. It ruled that the project would interfere with certain "existing uses" of the river by reducing dissolved oxygen levels and destroying an important stretch of river for fishing and boating.

Since there must be a section 401 certification for FERC to issue a license, the board's action appeared to stop the project. But a hydro project can have as many lives as a cat, and the company launched an effort in the Maine legislature to retroactively revise the state's water quality standards so that the project could proceed. The legislation easily passed the Maine House, but after an intensive lobbying effort by state conservation groups, it was defeated by a narrow vote in the Maine Senate.

In March 1986, after spending an estimated $6 million planning and promoting the project, Great Northern Paper Company announced that it was dropping the project. This important victory was achieved just as FERC began to evaluate the project.

The Comprehensive Planning Requirement

In addition to protecting the public interest, the Commission also has a responsibility, under section 10 of the Federal Power Act, to ensure that proposed hydroelectric projects conform to comprehensive river plans.[4] There are two aspects of the comprehensive planning provision. First, the Federal Power Act requires that the Commission itself, in the context of each project application, prepare some type of comprehensive plan for the use and development of the river. Second, as a result of ECPA, the Commission also has a duty to consider whether proposed hydroelectric projects are consistent with state and other federal comprehensive plans.

Chris Ayres

The West Branch of the Penobscot in Maine. This stretch of white water would have been flooded by the Big A dam.

The Commission's Plans

Section 10(a)(1) of the Federal Power Act defines the Commission's duty to prepare comprehensive river plans. It provides that the Commission can approve a development application only on the condition

> that the project adopted, including the maps, plans, and specifications, shall be such as in the judgment of the Commission will be best adapted to a comprehensive plan for improving or developing a waterway or waterways for the use or benefit of interstate or foreign commerce, for the improvement and utilization of water-power development, for the adequate protection, mitigation, and enhancement of fish and wildlife (including related spawning grounds and habitat), and for other beneficial public uses, including irrigation, flood control, water supply, and recreational and other purposes referred to in Section 4(e).

Under section 10(a)(1), the Commission must explicitly decide whether or not each proposed project conforms to a "comprehensive plan" for the river before issuing a project license.

The exact scope of the Commission's responsibilities has been controversial. Congress originally adopted the comprehensive planning requirement because it believed that individual hydroelectric proposals should be reviewed in the context of all other existing and potential uses of the entire river. When it adopted the requirement in 1920, Congress may well have been motivated, in part, by a desire that the hydroelectric potential of rivers be exploited to the fullest extent possible. However, in view of recent amendments to the Federal Power Act, the Commission's comprehensive plans must now achieve a balance between conservation and development within a particular river basin.

The Commission traditionally has taken the position that it can satisfy its comprehensive planning responsibilities by making its decisions on the basis of evidence presented at FERC proceedings. Based on a narrow reading of section 10(a)(1), the Commission has rejected repeated suggestions from conservationists that it has a responsibility itself to prepare written plans for river basins.[5]

In a series of recent decisions, however, the Court of Appeals for the Ninth Circuit has become increasingly critical of the Commission's failure to prepare comprehensive plans. In *LaFlamme v. FERC*[6] the court overturned a Commission license, in part because the Commission had failed to carry out its comprehensive planning responsibilities. The court observed that the "requirement of a comprehensive plan underscores Congress' commitment to coordinated study and comprehensive planning along an entire river system before hydroelectric projects are authorized." The court also stated that "each particular project should be given consideration not only with a view to the locality where it would be constructed but with reference to the entire system of which it constitutes a part." (See Fighting for the American on page 88 for a detailed description of the background to this important environmental case.)

Plans Prepared by Others

Section 10(a)(2)(A), which ECPA added to the Federal Power Act, requires for the first time that the Commission pay attention to the comprehensive plans prepared by the states and other federal agencies. It requires the Commission to consider

> the extent to which the project is consistent with a comprehensive plan (where one exists) for improving, developing, or conserving a waterway or waterways affected by the project that is prepared by—(1) an agency established pursuant to Federal law that has the authority to prepare such a plan, or (2) the State in which the facility is or will be located.

The legislative history of ECPA demonstrates that Congress wished to protect state rivers from inappropriate hydroelectric development. (Designation under the federal Wild and Scenic Rivers Act completely precludes hydroelectric development; the same is not true for designation under state laws unless the applicant is seeking PURPA benefits for a project involving a new dam or diversion.) Congress wanted the Commission to show greater deference than it had in the past to state and other federal agencies' river planning efforts.

In May 1988, the Commission issued its official interpretation of this comprehensive planning provision.[7] The FERC rule states that the Commission will recognize a plan under section 10(a)(2)(A) if (1) an authorized federal agency or an authorized agency of the state in which the facility would be located has prepared the plan; (2) the plan constitutes a "comprehensive study of one or more of the beneficial uses of a waterway or waterways;" (3) the plan includes a description of the standards applied, the data relied upon, and the methodology used in preparing the plan; and (4) the plan has been filed with the Commission.[8]

Under this broad interpretation of section 10(a)(2)(A), a wide variety of river-related plans have qualified. By April 1989, FERC had identified over 440 comprehensive plans. Since Congress specifically intended that state scenic river programs be included, they will certainly qualify. Other types of comprehensive plans include, for example, state water resource plans, state recreation plans, Forest Service and Bureau of Land Management resource management plans, anadromous fish restoration plans, and the Nationwide Rivers Inventory prepared by the National Park Service.

When a state or federal agency has a comprehensive plan that falls within the Commission's definition, how will the Commission carry out its responsibility to consider "the extent to which" the hydroelectric project is consistent with the plan? So far, there is no Commission or court precedent on this important question. Nonetheless, it is reasonable to conclude that the Commission should seldom, if ever, license a hydroelectric project on a river that a state or federal agency has specifically declared to be off-limits to development.

The state and federal planning process provides an important opportunity for river conservationists. The Commission will be required to give substantial deference to those policies included in state or federal agencies' comprehensive plans.

Miscellaneous Licensing Requirements

While the public interest test and the comprehensive planning requirement provide the basic framework for Commission decision making, other specific standards and requirements also apply.

Section 10 of the Federal Power Act, as recently amended by ECPA, establishes a specific mandate to protect fish and wildlife. The Commission must "adequately and equitably protect, mitigate damages to, and enhance, fish and wildlife (including related spawning grounds and habitat) affected by the development, operation, and management of the project."[9] Furthermore, this provision states that "each license *shall* include conditions for such protection, mitigation, and enhancement."[10] (Emphasis added.) Such conditions are generally required to be based on recommendations received from the National Marine Fisheries Service, the U.S. Fish and Wildlife Service, or state fish and wildlife agencies. (See Fish and Wildlife Agencies on page 53.)

Section 4(e) of the Federal Power Act states that the Commission can issue a license for a project within any federal "reservation" only after the Commission finds that the project "will not interfere or be inconsistent with the purpose for which such reservation was created or acquired." A federal reservation, as defined in the Federal Power Act, includes national forests, wildlife refuges and hatcheries, Indian reservations, military lands, and other similar lands carved out of the federal public domain, as well as federal lands that have been acquired and are being held for some public purpose.[11] This provision requires the Commission to consider, for example, whether a proposed project is consistent with Forest Service or Bureau of Land Management resource management plans and to reject the project if there is a clear conflict. (See chapter 7 for a discussion of the federal land management agencies' uncertain authority to rule on whether hydroelectric projects should be constructed on public lands under their management.)

As discussed earlier, inclusion of the phrase *energy conservation* in section 4(e) requires that energy conservation receive equal consideration as an alternative in the Commission's review of hydroelectric proposals. But another provision of revised section 10 imposes even more specific energy conservation obligations on public or private utilities (as opposed to independent power producers) engaged in hydroelectric development. Section 10(a)(2)(C) requires the Commission, in deciding whether or not to approve a development, to consider a utility's "electrical consumption efficiency improvement program . . . , including its plans, performance and capabilities for encouraging or assisting its customers to conserve electricity cost-effectively, taking into account the published policies, restrictions, and requirements of relevant State regulatory agencies."[12]

In other words, before approving a utility project, the Commission must specifically investigate the utility's available options for producing the same amount of power by encouraging more efficient electricity use by its customers. To implement this provision, the Commission wrote to utilities shortly after the enactment of ECPA, requesting that they file their energy conservation plans with FERC. These plans can be obtained by contacting the Commission's Office of Hydropower Licensing.

FERC'S RESPONSIBILITY TO CONSULT WITH OTHER ENTITIES

Apart from the substantive rules that govern the Commission's decision making, other provisions of the Federal Power Act require the Commission to consult with, and sometimes take the lead from, other entities that have special expertise—and established institutional interests—in protecting certain natural resources.

Fish and Wildlife Agencies

Mandatory Fish and Wildlife Conditions

Fish and wildlife agencies have authority to impose *mandatory* terms and conditions on certain types of projects. This means that FERC's license or exemption of such projects must include requirements developed by those agencies, whether or not the Commission approves of the requirements. Section 30(c) of the Federal Power Act requires the Commission, in acting on certain hydropower applications, (1) to consult with the Fish and Wildlife Service, the National Marine Fisheries Service, and the state fish and wildlife agencies for the state in which the facility is located and (2) to include in any order approving a project "such terms and conditions" as any of these agencies "determine are appropriate to prevent loss of, or damage to," fish and wildlife resources.[13]

The fish and wildlife agencies are authorized to exercise mandatory conditioning authority with respect to three types of projects:

1. conduit projects, which are exempt from the licensing requirements of the Federal Power Act and involve the installation of hydroelectric facilities at existing canals, ditches, or other constructions used to distribute water for agricultural, municipal, or industrial purposes[14]
2. small hydroelectric projects of less than five-megawatt capacity that are constructed at existing dams or that rely on natural water features[15]
3. hydroelectric projects involving the construction of new dams or diversions if the developers are seeking benefits under the Public Utility Regulatory Policies Act (discussed later in more detail)

The agencies have broad discretion to devise terms and conditions designed to protect fish and wildlife resources. Typical conditions require, for example, that a certain minimum stream flow be maintained immediately below a dam in order to preserve existing fish habitat or that upstream and downstream fish passage facilities be installed to support fish migration.

The Section 10(j) Process

The Commission has also a general mandate to consult with fish and wildlife agencies to obtain their recommended terms and conditions to protect fish and wildlife. The so-called section 10(j) process applies to all types of hydroelectric projects.[16] Unlike in the section 30(c) process, the Commission can reject section 10(j) recommendations from fish and wildlife agencies under certain circumstances.

Section 10(j) mandates that the Commission "adequately and equitably protect, mitigate damages to, and enhance, fish and wildlife (including related spawning grounds and habitat) affected by the development, operation, and management of the project." In addition, the Commission must include in each project license "conditions for such protection, mitigation, and enhancement."

Fish and wildlife conditions included in a Commission license "shall be based" on recommendations received from the National Marine Fisheries Service, the U.S. Fish and Wildlife Service, and state fish and wildlife agencies.[17] The Commission may determine, after reviewing the recommended conditions, that they are "inconsistent with the purposes and requirements" of the Federal Power Act or "other applicable law." In that case the Commission is required to attempt to resolve the conflict informally, "giving due weight to the recommendations, expertise, and statutory responsibilities of such agencies."[18]

If the Commission "does not adopt in whole or in part" a condition recommended by an agency, it is required to make each of the following *findings:*

- that adoption of the recommendations would be inconsistent with the purposes and requirements of the Federal Power Act or some other applicable law
- that the conditions selected by the Commission comply with its mandate to adequately and equitably protect, mitigate damages to, and enhance fish and wildlife (including related spawning grounds and habitat) affected by the development, operation, and management of the project[19]

If it intends to reject an agency's recommendations, the Commission must publish these findings together with a statement of the basis for each of the findings.

In practice, most disagreements between Commission staff and resource agency personnel over recommended terms and conditions are resolved informally, without resorting to the formal process for dispute resolution provided in the statute. Typically, staff members work out their differences over the telephone or through the mail. On some occasions, the Commission has convened formal meetings, sometimes involving the applicant and intervenors, to resolve section 10(j) disputes.[20]

Federal Land Management Agencies

Just as fish and wildlife agencies can impose mandatory terms on the licenses of certain hydroelectric projects under section 30(c), so too the Forest Service, the Bureau of Land Management, and the U.S. Fish and Wildlife Service can impose mandatory terms and conditions on projects located on certain federal lands. Under section 4(e) of the Federal Power Act, a license issued by the Commission for a project on a federal "reservation . . . shall be subject to and contain such conditions as the Secretary of the department under whose supervision such reservation falls shall deem necessary for the adequate protection and utilization of such reservation."[21]

For many years, the Commission claimed the authority to override conditions developed by the land management agencies. However, in *Escondido Mutual Water Co. v. LaJolla Band of Mission Indians,*[22] the U.S. Supreme Court affirmed that section 4(e) means what it says. The Court held that the Commission has an absolute duty to accept and implement conditions proposed by the Department of Agriculture or the Department of the Interior that relate to the protection and use of federal lands. The Court ruled that the Commission has no authority to reject or modify these conditions.

The Court suggested that there may be limits on the authority of federal land management agencies under section 4(e). "It may well be," the Court stated, "that in a particular case the conditions suggested by the Secretary [authorized to manage the reservation] will unduly undermine the Commission's licensing judgment." In other words, the Court suggested that the land management agencies may have authority under the Federal Power Act only to condition projects, not to veto them.

Apart from this authority under section 4(e), the land management agencies also appear to have independent regulatory authority over hydroelectric projects on federal lands, under the Federal Land Policy and Management Act (FLPMA). However, as discussed in chapter 7, the Commission and the land management agencies are currently disputing the scope of such authority under FLPMA.

Miscellaneous Consultation Requirements

As a result of ECPA, the Commission now has an explicit obligation to consider the recommendations of all federal and state agencies, as well as Indian tribes, that have jurisdiction over resources that may be affected by hydroelectric development. Specifically, section 10(a)(2)(B) of the Federal Power Act now requires the Commission to consider

> the recommendations of Federal and State agencies exercising administration over flood control, navigation, irrigation, recreation, cultural and other relevant resources of the State in which the project is located, and the recommendations (including fish and wildlife recommendations) of Indian tribes affected by the project.[23]

After receiving a hydropower application, FERC is required under section 10(a)(3) of the Federal Power Act to contact federal and state resource agencies and Indian tribes and request that they submit proposed terms and conditions for the project.[24] The Federal Power Act gives the Commission ultimate authority to decide whether or not to accept these proposed terms and conditions.

FERC'S RESPONSIBILITY TO STUDY ENVIRONMENTAL IMPACTS

Like all other federal agencies, the Commission is required to study the environmental effects of any proposed action under the National Environmental Policy Act (NEPA) before deciding whether or not to approve the action.[25] A complete discussion of the requirements of NEPA is beyond the scope of this handbook. However, there is certain basic information relating to Commission implementation of NEPA that you need to know.

NEPA is essentially a procedural law. It requires federal agencies to study and weigh the environmental consequences of proposed actions and to examine alternatives to those actions before proceeding. In addition, NEPA requires federal agencies to provide some opportunity for the public to comment on the proposed actions and their potential impacts. So long as a federal agency has complied with these procedural requirements, however, the agency has basically met its obligation under NEPA, whether or not the agency decides to proceed with the project. (Of course, the Commission probably would be violating other legal requirements, including the "equal consideration" mandate of ECPA, if it chose to ignore serious environmental problems disclosed by an environmental impact statement.)

In December 1987, following years of delay, the Commission adopted regulations that explain how it will apply NEPA to hydroelectric projects and other activities. A copy of the Commission's NEPA regulations is included in Appendix E.

The basic purposes of the regulations are twofold: to describe how the Commission selects the proper level of environmental analysis for particular projects and to explain how the public can participate in the NEPA process. The Council on Environmental Quality (CEQ), within the executive office of the president, also has published guidelines for compliance with NEPA.[26] The Commission regulations recognize that the CEQ guidelines apply to Commission activities, just as they do to any other federal agency.

Several NEPA issues repeatedly arise in hydroelectric proceedings. Each issue is discussed below.

Level of NEPA Analysis

The Commission has three options for analyzing how a proposed project would affect a river. These are: (1) an environmental impact statement (EIS), a detailed study of potential environmental impacts of and alternatives to the proposed action; (2) an environmental assessment (EA), an abbreviated version of an environmental impact statement; and (3) a categorical exclusion, that is, no environmental analysis. A categorical exclusion is based on the Commission's determination from past experience that an action of that type produces no significant environmental impacts. An EA is sometimes prepared first to determine whether an EIS is necessary. If the Commission concludes, based on the EA, that a project may "significantly [affect] the quality of the human environment,"[27] it must proceed with an EIS.

A great deal can turn on the level of environmental analysis the Commission conducts. It will examine the environmental issues most thoroughly if it prepares a full EIS. The analysis and study will be less thorough if the Commission prepares only an EA. And it will not consider environmental effects at all if it decides the action is categorically excluded from the Commission's NEPA obligation. In addition, there generally will be opportunities for public involvement in the NEPA process only if the Commission prepares an EIS.

Because of the expense involved, the Commission has prepared fewer than 50 environmental impact statements since the early 1970s. Based on the low budget commitments, FERC is planning to prepare only a handful each year for the foreseeable future. Especially given the large number of relicensing proceedings underway or scheduled to commence in the near future (approximately 175 project licenses will expire in 1993), Congress and the Commission probably have not allocated sufficient resources to environmental analysis.

FERC ultimately must decide on a case-by-case basis whether an EIS (or an EA) is required. The Commission's NEPA regulations state that it will decide what level of analysis is needed on the basis of the facts for each individual project or action.[28] Nonetheless, to guide the FERC staff as well as the public, the Commission's NEPA regulations indicate the level of environmental analysis—that is, categorical exclusion, EA, or EIS—that *normally* will be appropriate for different types of projects or actions. This part of the Commission's NEPA regulations is summarized in the table below.

FERC Guidelines on Appropriate Level of NEPA Analysis for Different Types of Projects and Commission Actions

Environmental Impact Statements
- Issuance of license for construction of new hydroelectric projects

Environmental Assessment
- Issuance of license for hydroelectric project at existing dam
- Issuance of exemption for project at existing dam, including project that relies on natural water features, with generating capacity of less than five megawatts
- Issuance of license for additional project work at already licensed project
- Issuance of new license for existing project, i.e., relicensing
- Surrender of water power license or exemption where project work exists or ground-disturbing activity has occurred
- Amendment to license or exemption that requires ground-disturbing activity or changes to project work or operation

Categorical Exclusion
- Transfer of license or exemption to new owner
- Issuance of preliminary permit
- Surrender or amendment of preliminary permit, and surrender of license or exemption, where no project work exists and no ground-disturbing activity has occurred
- Amendment to license or exemption that does not require ground-disturbing activity or changes to project work or operation
- Issuance of exemption for conduit hydroelectric project

NOTE: These are only guidelines. The Commission's NEPA regulations expressly indicate that FERC will perform a more detailed environmental analysis than these guidelines call for if the facts indicate that it would be appropriate.

Cumulative Environmental Impacts

Another potentially important issue is that of the cumulative effects of a proposed project in light of the impacts of other development in the same river basin. Cumulative impacts analysis is important, for example, in determining whether yet another dam on a river would defeat ongoing efforts to restore or enhance fisheries. Similarly, cumulative impacts analysis of existing dams scheduled for relicensing may help describe the existing quality of the habitat and define appropriate mitigation measures. Cumulative impacts analysis is useful also for determining whether a proposed project would produce a fair and reasonable balance between power generation and white-water boating.

Under NEPA the Commission has a clear obligation to consider cumulative impacts. The guidelines of the Council on Environmental Quality, which apply to the Commission, define a cumulative impact as "the impact on the environment which results from the incremental impact of the action when added to other past, present, and reasonably foreseeable future actions."[29] The Commission must consider a proposed project in light of the present impacts of existing hydroelectric and other developments and the future impacts of reasonably foreseeable projects. The Commission's adoption of new NEPA regulations should produce more thorough and objective Commission analysis of cumulative impacts. Conservationists are continuing in their attempts to push the Commission in that direction.

Alternatives

A final issue is whether the Commission's environmental analysis considers an adequate range of alternatives to the proposed project. Too often, the Commission has focused solely on whether or not to approve a project as proposed, without investigating options not presented by the developer that could meet the goals of the project at less environmental cost. For example, it is almost always worthwhile in evaluating a hydroelectric project to consider whether the same or more electric power could be produced in other, less environmentally damaging ways.

ENVIRONMENTAL STANDARDS FOR PURPA BENEFITS

As discussed in chapter 1, the Commission's controversial decision to extend PURPA benefits to projects involving new dams and diversions has had a major role in spurring hydropower development of free-flowing rivers. In ECPA, Congress began to address this controversy by adopting a moratorium on PURPA benefits for new dam and diversion projects, with certain exceptions for "grandfathered" projects (described in more detail later). The moratorium allows the Commission and Congress to assess whether PURPA benefits should go to new dam and diversion projects. Section 8(d) of ECPA directed the Commission to conduct a study and submit recommendations to Congress. It also provided a deadline. The moratorium will remain in place from October 1986 through the first full session of Congress after the report and recommendations have been submitted to Congress. Based on the current schedule of the Commission, it appears that the moratorium will expire at the end of 1989.

At the time the moratorium was adopted, Congress also established three stringent requirements for new dam and diversion projects.[30] These new requirements should direct PURPA developers away from outstanding rivers. Of course, these new requirements do not apply to developers, such as utilities, that do not rely on PURPA. If the moratorium ends and Congress has not further amended PURPA, a new dam or diversion project will be able to receive PURPA benefits only if it meets the following three requirements.

1. *No substantial adverse effects.* A new dam or diversion project is not eligible for PURPA benefits unless the Commission determines, when it decides whether to issue a license for the project, that it "will not have substantial adverse effects on the environment, including recreation and water quality." Recently, the Commission defined *substantial adverse effects* to mean "a substantial alteration in the existing or potential use of, or loss of, natural features, existing habitat, recreational uses, water quality, or other environmental resources."[31]
2. *Not located on a protected river.* The "protected rivers" requirement bars PURPA benefits for a new dam or diversion project if, at the time the application is accepted for filing, the Commission finds that the project would be located on either one of two categories of rivers.

   The first category consists of river segments designated for potential inclusion in state or federal *wild and scenic river systems* or already included in state wild and scenic river systems. (As discussed in chapter 7, a river already designated as a federal wild and scenic area is protected from new hydroelectric development, whether or not the developer seeks PURPA benefits.) In our view, a river is designated for potential inclusion by a state if the state is actually studying the river for official designation or if the river has been identified as being worthy of protection in an official state river assessment or river inventory. At a minimum, federally "designated" rivers should include those currently under study for inclusion in the federal system and river segments identified in the National Park Service's 1982 Nationwide Rivers Inventory as being worthy of consideration for inclusion in the system.

   The second category of protected rivers includes rivers that the states have determined possess "unique natural, recreational, cultural or scenic attributes which could be adversely affected by hydroelectric development." The Commission recently defined this provision as follows:

> a river is protected if a state, either by or pursuant to an act of the state legislature, has determined that the watercourse possesses unique natural, recreational, cultural, or scenic attributes that would be adversely affected by hydroelectric development. This . . . means that . . . the state could enact a specific statute to protect a specific river, or . . . the state or a political subdivision thereof (e.g., a department of wildlife conservation, a department of environmental conservation, a fish and wildlife department, etc.) could, pursuant to a statute, designate a river for protected status under state law.[32]

3. *Compliance with fish and wildlife conditions.* This provision bars PURPA benefits for new dams or diversions unless the projects are constructed in accordance with mandatory conditions set by federal and state fish and wildlife agencies.

A project that fails to meet any of these requirements is disqualified only from receiving PURPA benefits. These benefits are generally critical to the success of a small project. But developers could construct projects without complying with any of these requirements so long as they do not seek PURPA benefits.

As mentioned above, ECPA grandfathered certain projects from the moratorium under specified conditions. These include projects for which applications were filed with the Commission prior to enactment of ECPA (16 October 1986) and accepted for filing within the following three years. These projects are exempt from the moratorium so long as they are not inconsistent with the protected rivers requirement.[33] Other projects are exempt from the moratorium if the applicants can demonstrate that "substantial monetary resources" were committed to the projects prior to the adoption of ECPA and if the projects satisfy the protected rivers and no substantial adverse effects requirements.[34] Any applicant who wished to rely on this provision was required to file with the Commission, by April 1988, a petition demonstrating that the substantial monetary commitment test had been met.

REFERENCES

1. 16 U.S.C. § 797(e).
2. 387 U.S. 428 (1967).
3. 387 U.S. at 450.

4. 16 U.S.C. § 803(a).
5. For a recent Commission interpretation of its comprehensive planning responsibilities, see *In re City of Fort Smith*, 44 FERC ¶ 61,160 (CCH) (28 July 1988).
6. 842 F.2d 1063 (9th Cir. 1988).
7. Order No. 481, 52 Fed. Reg. 39905 (26 Oct. 1987).
8. 53 Fed. Reg. 15802 (4 May 1988). The new rule appears in the Commission's regulations at 18 C.F.R. § 2.19.
9. 16 U.S.C. § 803(j)(1).
10. 16 U.S.C. § 803(j)(1).
11. 16 U.S.C. § 796(2).
12. 16 U.S.C. § 803(a)(2)(C).
13. 16 U.S.C. § 823a(c).
14. 16 U.S.C. § 823a.
15. 16 U.S.C. § 2705.
16. 16 U.S.C. § 803(j).
17. 16 U.S.C. § 803(j)(1).
18. 16 U.S.C. § 803(j)(2).
19. 16 U.S.C. § 803(j)(2).
20. At a meeting in January 1989, the Commission publicly announced that it was going to initiate a rule-making proceeding to develop formal regulations governing the section 10(j) process.
21. 16 U.S.C. § 797(e).
22. 466 U.S. 765 (1984).
23. 16 U.S.C. § 803(a)(2)(b).
24. 16 U.S.C. § 803(a)(3).
25. 42 U.S.C. §§ 4321 et seq.
26. 40 C.F.R. §§ 1500.1–1507.7.
27. 42 U.S.C. § 4321.
28. See 16 C.F.R. §§ 380.4(b) and 380.5(a).
29. 40 C.F.R. § 1508.7.
30. See 16 U.S.C. § 824a-3(j).
31. Order No. 499, 53 Fed. Reg. 26992 (11 July 1988).
32. 53 Fed. Reg. 40722 (18 Oct. 1988).
33. See 16 U.S.C. § 824a-3 note.
34. See 16 U.S.C. § 824a-3 note.

Chapter 4

Relicensing: Standards and Procedures

FERC's relicensing standards • Intervention procedures • License denial, federal takeover, and nonpower license as alternatives

SINCE IT WAS ESTABLISHED in 1920, the Commission has focused on proposals to construct new hydroelectric facilities. That will change over the next five years as the Commission turns increasing attention to the *relicensing* of existing hydroelectric projects. The Commission must decide whether these projects should continue to be operated as hydropower facilities and, if so, by whom and in accordance with what new terms and conditions.

The public can and should participate in the relicensing process. Indeed, if the public does not participate, business is likely to proceed as usual, and an important opportunity to reclaim some of America's working rivers may be lost. About 275 projects—175 in 1993 alone—will come up for relicensing by the end of the next decade. Relicensing of each project will consume several years, and it is important that the public get involved in the process at the earliest stages.

AN OVERVIEW

Congress adopted the Federal Water Power Act of 1920 to serve several conflicting objectives. Congress wished to encourage private investment in the construction of hydroelectric projects. Congress also wanted to ensure that, at least over the long term, the public would retain ultimate control over the use of river resources.

Congress resolved these conflicting objectives by authorizing the Commission to issue private development licenses that have a maximum life of 50 years. Congress believed that a 50-year license provided a sufficient period of time for an investor to amortize the costs of constructing a project and to receive some financial reward for having put capital at risk. A term of 50 years provided a specific deadline by which the license would expire and the public would have an opportunity to affirmatively decide upon future uses of the river resource.

Relicensing of an existing hydroelectric facility raises different issues from those raised in the licensing of a new project. Since the project already exists, the natural character of the river obviously has been altered by a dam, a diversion, an impoundment, or other project works. While the natural character of the river has probably been degraded, certain river uses may actually have been enhanced by the project. For example, white-water boaters may have come to depend upon periodic releases from a hydroelectric dam, or a specific type of

Lycoming Creek in Pennsylvania. Diversion and dewatering of streams can seriously impact fishing opportunities. Hydroelectric relicensing provides an opportunity to obtain improved flows.

fishery may have been adapted to the water flow below a dam. In addition, the project already is operating, and the owner and power customers almost certainly have come to depend upon the project as a source of electricity.

In most instances, older hydroelectric facilities that are coming up for relicensing were constructed with little or no regard for environmental values. All of the projects were constructed prior to the enactment of the National Environmental Policy Act (NEPA), and many were constructed prior to the enactment of the Federal Power Act, when there was no public process for reviewing the project proposals. Relicensing provides an opportunity to conduct a proper environmental review of a completed project and to change its structure or operation to protect, mitigate, and enhance environmental and recreational values. Since the owner already has had the opportunity to fully amortize the investment in the project, the owner has no right to expect that business will continue as usual. In a sense, relicensing involves the return of river resources to the public. The Commission must decide how the project should be managed in the future to best serve the public interest.

Relicensing provides an opportunity for dam owners to implement

John D. Echeverria

The Pine River in Wisconsin. Local paddlers hope to use the relicensing process to get weekend releases in this dewatered section of river.

comprehensive plans developed by state or federal agencies. Some relatively common relicensing requirements that benefit fisheries include installation of fish passage facilities, so that migratory fish can circumvent a dam, or new controls on the amount and timing of flow releases below a dam, so that dewatered stretches of river can once again be productive. Appropriate relicensing requirements that benefit recreational boating might include new or improved access sites for boaters or modified downstream releases that provide better flow levels for boating on weekends or at other preferred times of the week.

The great majority of hydroelectric projects whose licenses expire over the next decade probably will be relicensed as operating hydroelectric projects. At the same time, there are specific instances—arising out of a project's poor financial return, safety problems, serious environmental degradation, or a combination of these factors—when the public interest will be served best by converting the site to another use, perhaps by removing the dam altogether. As discussed in Alternatives to Relicensing, beginning on page 71, the Federal Power Act provides several specific procedural options for converting existing hydroelectric projects to nonpower uses.

FERC'S RELICENSING STANDARDS

In deciding whether or on what conditions to issue a new license for an existing project, the Commission is bound by the same standards that apply to applications for new projects.[1] Thus, under section 4(e) of the Federal Power Act, the Commission must determine what is "desirable and justified in the public interest."[2] The Commission must give "equal consideration" to "power and development purposes" and to "the purposes of energy conservation, the protection, mitigation of damage to, and enhancement of, fish and wildlife (including related spawning grounds and habitat), the protection of recreational opportunities, and the preservation of other aspects of environmental quality."[3] In addition, under section 10 of the Federal Power Act, the Commission must ensure that the project is "best adapted" to a comprehensive river plan, and it must consider the extent to which issuing a new license for an existing project would be consistent with a comprehensive plan prepared by the state or another federal agency.[4]

Furthermore, relicensing applications are subject to the Electric Consumers Protection Act (ECPA). The act requires the Commission to consider the recommendations of federal and state agencies, as well as Indian tribes, that have jurisdiction over resources that may be affected by hydroelectric development and to solicit proposed terms and conditions from these agencies and tribes.[5] Relicensing applications must include conditions "to adequately and equitably" protect fish and wildlife resources, and the conditions generally must be based on the recommendations received from fish and wildlife agencies.[6] Finally, Commission review of hydroelectric projects must be conducted in accordance with the requirements of NEPA.

In addition to these basic standards and procedures, however, the Federal Power Act establishes additional standards specific to the relicensing decision. Section 15 of the act, as amended by ECPA, directs the Commission to consider the following in reviewing the application of the incumbent and any competitor.[7]

1. the plans and abilities of the applicant to comply with the act and the conditions of the license
2. the safety of the applicant's proposal
3. the plans and abilities of the applicant to provide efficient and reliable electric service
4. the applicant's need for power
5. the applicant's existing and planned transmission services
6. the cost-effectiveness of the applicant's plans
7. such other factors as the Commission deems relevant

In addition, in considering the application of an incumbent for a new license, the Commission must consider (1) the incumbent's record of compliance with the terms and conditions of the existing license and (2) any "actions taken by the existing licensee related to the project which affect the public."[8] These criteria are explained in greater detail in formal relicensing regulations that the Commission issued just as this book was going to press.[9]

Prior to the enactment of ECPA, the Federal Power Act authorized the Commission to relicense a project by issuing the license to either a new licensee or the original licensee. At least as between private applicants, the act gave the incumbent no advantage. (It did require the new licensee to return the original licensee's "net investment" in a project plus so-called severance damages, a combined amount that is likely to be far less than the market value of a project.) As between a private applicant and a municipal utility, it was debatable whether the so-called municipal preference, which applies in original licensing proceedings, also applied in relicensing.

The Commission's experience with relicensings in the late 1970s demonstrated that the threat of project transfer stimulated incumbent licensees and new applicants to outdo each other in developing new, environmentally beneficial project terms and conditions. Incumbent licensees raised major objections to a competitive process. They argued that new licensees would receive the projects by paying substantially less than they were worth and that numerous project transfers would disrupt the present system of generating and transmitting electric power.

In ECPA Congress granted incumbents an advantage—but only a small advantage—over new applicants in the relicensing process. Congress made clear that the municipal preference would not apply in relicensing proceedings. Congress directed the Commission to grant a license to "the applicant having the final proposal which the Commission determines is best adapted to serve the public interest." The Commission is required to ensure that "insignificant differences" between applications do not result in the transfer of a project to a new owner.

The legislative history of ECPA grants the Commission fairly broad latitude to interpret "insignificant differences" so as to encourage or discourage competition. The legislative history suggests that the Commission should examine the incumbent's record of compliance with the terms of the original license in determining whether to grant the license to a new applicant. As stated by one Congressman, "If there are only insignificant differences, then the existing licensee's track record will be dispositive."[10]

There is one important exception to the general rule that there should be open competition in the relicensing process. The Federal Power Act, as amended by ECPA, states that terms and conditions for the protection of fish and wildlife shall be determined through the same section 10(j) process that applies in original licensing proceedings. The act specifically directs the Commission not to develop fish and wildlife conditions through "a comparative evaluation."[11] This provision reflects Congress's determination that the section 10(j) process should provide adequate protection for fish and wildlife resources, and a desire to avoid what Congress considered "gold plating" of existing projects for the benefit of fish and wildlife.

The absence of competition over fish and wildlife terms will discourage somewhat the development of environmentally beneficial relicensing proposals. On the other hand, this anti-gold-plating provision applies only to fish and wildlife conditions. Applicants can still compete based on their proposed projects' impacts on recreation or scenic values, for example.

PROCEDURAL ASPECTS OF INTERVENING IN RELICENSING PROCEEDINGS

The procedures governing the relicensing process are similar to those governing original project applications, with special provisions to accommodate the unique features of relicensing. At least five years before the expiration of a license, the incumbent licensee is required to inform the Commission of the intention to file or not to file an application for a new license.[12] The purpose of this provision is to provide a starting point for the relicensing process so that the Commission ordinarily will be in a position to issue a license by the time the original license expires.[13]

The Commission publishes in the *Federal Register* and local newspapers a notice of the incumbent licensee's intent to file or not to file. In addition, the Commission is required to specifically inform federal and state resource agencies of the licensee's decision.[14]

At the same time that the licensee files a *notice of intent* to seek relicensing, the licensee must make available for public inspection at the company's offices a variety of information concerning the project. This includes, for example, information regarding the project's generation, finances, and environmental effects.[15] The incumbent licensee must make copies of the required information available at reasonable cost and must open the project files to the public even if the licensee is not seeking a new license. Importantly, what must be made available for public inspection relates only to existing operations. It does not include the incumbent's proposals for the new conditions to be included in the new license. Nor does it include a description of the environmental studies the incumbent intends to perform to support the relicensing application.

The Commission has adopted a rule providing guidance on how the public disclosure process should operate.[16] The rule enumerates in detail the types of information that a licensee must make available for inspection and copying. If a person believes that a licensee is not making the necessary information reasonably available, the rule provides a procedure for filing a petition with the Commission to resolve the dispute.[17]

The deadline for filing a relicensing application is two years prior to the expiration of the original license.[18] During the period between notification of intent to file and actual filing, the incumbent licensee is expected to proceed through the same three-stage consultation process that applies to original license applications (see Consultation with Other Agencies on page 52). This process is a crucial part of relicensing, in which the resource agencies identify their environmental concerns about a project, request appropriate studies, and attempt to assist the licensee in developing an acceptable application for a new license.

As this book was going to press, the Commission promulgated final relicensing regulations that also provide a limited opportunity for the public to participate in the consultation process. The applicant must invite the public to participate in the first stage of the process or, if the first stage already has been completed, in a special public meeting to review the applicant's relicensing plans.

Once the Commission has reviewed the application and accepted it for filing, the Commission publishes a public notice of the application in the *Federal Register* and local newspapers. The notice sets a *comment date,* which represents the deadline for filing motions to intervene.

Thereafter, the Commission processes a relicensing application in much the same fashion as applications for new projects. The Commission routinely prepares environmental assessments on relicensing applications. It will likely prepare environmental impact statements only for a major or particularly controversial project. The Commission also may conduct an adjudicatory hearing on the relicensing application if a party requests a hearing and if material factual questions about a project remain unresolved.

Under ECPA, the Commission must proceed expeditiously with relicensings. Within 60 days of the two-year deadline for submitting an application, the Commission must publish a notice establishing "expeditious procedures for relicensing."[19] Despite this statutory mandate, it is almost inevitable that the Commission will fail to complete processing of some relicensing applications by the time the original licenses expire. In that event, FERC may issue an annual license, which will generally extend the terms of the original license for the project until a new license is issued.

The Commission has discretion in setting the term of the new license, except that the license must be for no less than 30 years and no more than 50 years.[20] The Commission intends to issue short-term licenses for applications that involve no major project changes, and to grant longer-term licenses for applications that involve substantial project reconstruction or improvement.

ALTERNATIVES TO ISSUANCE OF A NEW POWER LICENSE

While it undoubtedly will relicense most existing projects, the Commission has other alternatives. These are: *denial* of a new hydroelectric license, *federal takeover,* and a *nonpower license.*

Several attempts have been made to obtain a nonpower license or to accomplish a federal takeover, but none has yet succeeded. Also, so far as we know, the Commission has never denied a new license application to a dam owner who wanted to continue operating the facility for hydropower. Nevertheless, a discussion of the options follows.

Denial of a New License

In some instances, the public interest may be served best by either removing an existing dam or converting it to a nonpower use. The proposal of Secretary of the Interior Hodel in late 1987 to remove Hetch Hetchy Dam from Yosemite National Park is a dramatic reminder that a dammed river does not necessarily have to stay that way forever.

So far as we know, the Commission has not yet denied a new license to any relicensing applicant. The Commission has authority to do so. Section 23b of the Federal Power Act makes it "unlawful" for any person to "operate or maintain" a hydroelectric project except in accordance with a valid Commission license. Read literally, this provision would prohibit continued operation of a project if the original license expired and if the Commission did not issue a new license. The only court decision that even addresses the issue indirectly, *Lac Court Oreilles Band of Lake Superior Chippewa Indians v. FPC*,[21] strongly supports the conclusion that the Commission may refuse to issue a new license for a project.

A practical obstacle to getting the Commission to deny a relicensing application is its concern about maintenance of the project facilities once the original license expires. A license, including even an annual license, provides the Commission control over a project, permitting it to ensure the project is safely maintained. The Commission's jurisdiction to regulate the project might terminate when the license is not renewed. The Commission might be reluctant to rely on a disappointed relicensing applicant to remove project facilities, even if it legally has the power to require such action.

In deciding whether to grant or deny a new license for a project, the Commission applies the same standard that governs all licensing decisions—what will best serve the public interest? Denying a relicensing application would eliminate an existing generating source of electricity. To make a strong case for denying an application, you would need to demonstrate that other existing and planned power sources would be sufficient to meet expected demand, and that continued use of the site for power generation would conflict with other uses, such as restoration of a spectacularly scenic valley or a stretch of river valuable for fishing or boating.

Federal Takeover

Another alternative to relicensing is federal takeover, which can occur upon the expiration of the original license if Congress decides that the project should be federally owned. The United States can then operate the project for power generation or devote the project to some other use.[22] For example, it may use the project as a recreational reservoir or restore the free-flowing character of the river. No federal takeover has yet occurred.[23]

The United States would have to compensate the original licensee for a takeover. Specifically, it would have to pay the licensee's "net invest-

American Rivers (furnished by Central Nebraska Public Power and Irrigation District)

Kingsley Dam on the Platte River in Nebraska. Reduced flows downstream from the dam threaten habitat for endangered whooping cranes and other birds.

ment" up to the "fair value of the property taken."[24] Net investment is calculated by taking the original cost of a project, increasing it by the cost of subsequent capital improvements, if any, and reducing it by the total accumulated depreciation. This sum may be a small fraction of the current value of a project as determined by the value of the electricity it is currently generating.

In addition to net investment, the project owner would be entitled to receive "severance damages," which would compensate for any injury to the owner's electricity-generating equipment that would be caused by disconnection of the project from the other equipment. Severance damages would be relatively inconsequential.[25]

Upon takeover, the United States would "assume all contracts entered into by the licensee with the approval of the Commission."[26] Under section 22 of the act, a licensee is prohibited from entering into a power sale contract extending beyond the termination date of the outstanding license without the approval of the Commission.[27] Thus, the Commission has the power to bar a contract which the United States would subsequently have to assume if the project were taken over.

Prior to 1968, the Commission submitted a report to Congress for each project subject to relicensing. In each report the Commission made a recommendation regarding federal takeover. The Commission then proceeded with the relicensing when Congress decided not to take over the project. In 1968, Congress strengthened the presumption in favor of relicensing.[28] Now, the Commission considers the alternative of recommending a federal takeover in the course of the relicensing proceeding. Congress becomes involved only if the Commission or some other agency makes an affirmative recommendation in favor of a takeover.[29]

If the Commission decides to recommend a federal takeover, it forwards the recommendation to Congress.[30] The Commission may *stay* the effective date of a new license while Congress reviews the takeover recommendation for the project, and it must issue such a stay upon the motion of any federal department or agency. Congress must decide whether to enact legislation authorizing a takeover and appropriating any funds that are needed to compensate the original licensee. If the original license expires while a takeover recommendation is pending before Congress, the Commission must issue one or more annual licenses to the original licensee until the project is taken over or the Commission issues a new license.

Nonpower License

The final alternative to relicensing is issuance of a nonpower license for the project.[31]

The Commission can issue a nonpower license so that the site of an existing project may be restored to its natural condition or so that a governmental entity may manage the project facilities for a nonpower purpose, such as recreation. A nonpower license is intended to be short-term,[32] lasting only until the site has been restored or until another governmental agency has agreed to assume jurisdiction over the project facilities.[33]

Unlike a takeover, which can be accomplished only by the federal government (though it may be recommended by a private party), virtually any person or entity, public or private, may apply for a nonpower license. The Federal Power Act authorizes the Commission to issue a nonpower license on its own motion or upon the application "of any licensee, person, State, municipality, or State [utility] Commission."[34]

To issue a nonpower license, the Commission must find that, "in conformity with a comprehensive plan for improving or developing a waterway or waterways for beneficial public uses[,] all or part of any licensed project should no longer be used or adapted for power purposes."[35] The act's legislative history, the Commission's orders, and court decisions provide little guidance as to the purposes and limitations of a nonpower license. Congress apparently believed that a nonpower license would be issued when the original licensee was unable or unwilling to continue to operate the project.[36] However, the Commission can also issue a nonpower license whenever a nonpower use of a project is best suited to "a comprehensive plan for improving or developing" a river, even if the incumbent licensee would prefer to continue to operate the project for power purposes.

This conclusion is supported by *Confederated Tribes and Bands of the Yakima Indian Nation v. FERC*,[37] in which the Court of Appeals for the Ninth Circuit stated: "Seemingly, the Commission would determine that a nonpower license is necessary if it concluded that power production needs were outweighed by recreational or environmental concerns." The court held that the Commission had to prepare an environmental impact statement in that relicensing proceeding, to provide the information necessary to decide whether a nonpower license should be issued for the project. In that case, the court ruled that proper consideration of the nonpower option "necessitate[d] the information prepared in an environmental impact statement."[38]

An applicant for a nonpower license must develop and file an application in much the same way as an applicant for a power license. For example, the Commission's proposed relicensing regulations provide that a nonpower applicant, like any other applicant, must go through the three-stage consultation process described in chapter 2.[39] Also, a nonpower applicant is entitled, as party to the consultation process, to gain access to the project site to gather information and conduct site visits with resource agencies. The requirement to go through the consultation process means that the person or group considering filing for a nonpower license should start the process at least three or four years prior to the expiration of the original license.

Like an applicant for a power license, an applicant for a nonpower license must file an extensive application. The application must include a description of the proposed nonpower use, a statement on how converting the project to a nonpower use would affect power supplies, an estimation of the costs of converting the project to a nonpower use, and an explanation of why the proposal should be granted. The Commission must provide public notice of the application, and federal and state agencies and any interested person or group has an opportunity to comment or to intervene in the proceeding.

REFERENCES

1. 16 U.S.C. § 808(a)(2).
2. 16 U.S.C. § 797(e).
3. 16 U.S.C. § 797(e).
4. 16 U.S.C. § 803(a).
5. 16 U.S.C. § 803(a)(2)(B).
6. 16 U.S.C. § 803(j).
7. 16 U.S.C. § 808(a)(2).
8. 16 U.S.C. § 808(a)(3).
9. See Commission Order No. 513, 54 Fed. Reg. 23756 (2 June 1989).
10. 123 Cong. Rec. H8950 (daily ed. 2 Oct. 1986).
11. 16 U.S.C. § 808(a)(2)(G).
12. 16 U.S.C. § 808(b)(1).
13. When Congress added the five-year notice provision to the Federal Power Act, certain outstanding licenses had less than five years to run. Accordingly, ECPA authorized the Commission to adjust the time period for filing a notice of intent when the developer had insufficient time to file a five-year notice before the license expired. See 16 U.S.C. § 808(c)(2).
14. 16 U.S.C. § 808(b)(3).
15. 16 U.S.C. § 808(b)(2).
16. 18 C.F.R. §§ 16.15–16.16.
17. 18 C.F.R. § 16.16(f).
18. 16 U.S.C. § 808(c)(1).

19. 16 U.S.C. § 808(c)(1).
20. 16 U.S.C. § 808(e).
21. 510 F.2d 198 (D.C. Cir. 1975).
22. 16 U.S.C. § 807.
23. Unsuccessful federal takeover attempts are discussed in *Lac Court Oreilles Band of Lake Superior Chippewa Indians v. FPC*, 510 F.2d 198 (D.C. Cir. 1975) and *In re Escondido Mutual Water Co.*, 6 FERC ¶ 61,189 (26 Feb. 1979).
24. 16 U.S.C. § 807(a).
25. See *In re Pacific Power and Light*, 23 FERC ¶ 63,037 (CCH) (28 April 1983) and *In re Escondido Mutual Water Co.*, 9 FERC ¶ 61,241 (CCH) (26 Nov. 1979).
26. 16 U.S.C. § 807(a).
27. 16 U.S.C. § 815.
28. Pub. L. No. 90–451.
29. 16 U.S.C. § 807(b).
30. 16 U.S.C. § 807(b).
31. 16 U.S.C. § 808(f).
32. 16 U.S.C. § 808(f) (a nonpower license is a "temporary" license).
33. The Commission is supposed to terminate its jurisdiction over a former power project. As soon as "a State, municipality, interstate agency, or another Federal agency is authorized and willing to assume regulatory control and supervision of the land and facilities included under the nonpower license and does so." 16 U.S.C. § 808(f).
34. 16 U.S.C. § 808(f).
35. 16 U.S.C. § 808(f).
36. For the legislative history of Public Law 90–451, which added the nonpower license provision to the Federal Power Act, see 1968 U.S. CODE CONG. AND AD. NEWS 3081.
37. 746 F.2d 466, 476 (9th Cir. 1984).
38. 746 F.2d at 476.
39. See FERC Order No. 513 (17 May 1989), to be published in 18 C.F.R. Part 16.

Chapter 5

Participating in the FERC Process

Learning about a project number, manager, status • The basics of intervening • Petitions for appeal or rehearing • Postlicensing consultations and complaints

WHAT DO YOU DO if your favorite trout stream is threatened by a hydroelectric proposal, or if you would like to see environmental improvements at an existing dam? You have to become a participant in the regulatory process. This chapter explains the basic steps you need to take to participate in FERC's review of a hydropower proposal.

LEARNING ABOUT A PROJECT

Citizens become aware of hydropower proposals in a variety of different ways. You might read a public notice of the application in a newspaper. A neighbor might mention a rumor about a proposed facility. You may fish for years below a particular dam and wonder how the amount of water in the stream could be increased. Or you may receive a notice in the mail that your land is included within the boundaries of a proposed hydroelectric project!

How do you find out what is going on? Here are some suggestions.

Getting the Project Number

A key piece of information is the project number that FERC assigns to the project. The project number usually has four or five digits—for example, project no. 7987. A specific FERC proceeding relating to a particular project—such as a preliminary permit proceeding, a licensing proceeding, or an amendment proceeding—may be identified by additional digits: project no. 7987–003.

You must have the project number to obtain information about a project from the FERC staff. The project number is useful also in communicating with other federal or state officials about the regulatory proceeding.

The FERC number should not be difficult to locate. The public notice of the application in the newspaper, or any official correspondence relating to the project, will include the project number. The staff of the Office of Hydropower Licensing at FERC may be able to help you locate the project number if you know the name of a project on a given stream. State and federal resource personnel reviewing the project application also can give you the project number.

Talking with the FERC Project Manager

If you want a status report on the project, you can contact the FERC project manager, the staff person assigned to process the application. You can get the person's name by calling the Office of Hydropower Licensing and telling them the project number. The single most important thing to determine from the FERC staff is whether a deadline has been set for filing a motion to intervene.

Getting on the FERC Service List

The best way to keep informed about new developments relating to a specific project is to be included on the official *service list.* As an intervenor, you will be included automatically on the service list and will receive every submittal to FERC and notice of every action by FERC, relating to that project. Even if you are not an intervenor, you may write to the secretary of the Commission and request to be included on the service list. (See Appendix A for FERC's address.)

Talking with the State Hydropower Coordinator

You also can learn about a project by contacting the staff in the state resources agency that deals with hydropower. In many states a specific person has been appointed as "hydro coordinator."

Getting the FERC Records

There is no substitute for reading the actual FERC file on a particular project. A FERC file may be voluminous, so it can take some time for the Commission staff to locate the specific records you want.

A useful first step is to get the *docket sheet,* a chronological listing of all the filings and orders in a particular proceeding. You can get the docket sheet by ordering it in person or by letter from the Commission's public information room (located on the first floor of the FERC building at 825 North Capitol Street). If you make your request by mail, include your daytime telephone number. The Commission staff will telephone you and inform you of the charge, if any, for obtaining the docket sheet. If the project is less than five years old, the staff can generally give you a copy of the docket sheet within an hour. If you are interested in an older project, you may have to wait a day or so. You cannot order a docket sheet over the telephone.

Once you review the docket sheet, you should be able to identify which documents will be relevant to your interests. You can obtain copies of the documents, like the docket sheet, by requesting them in person or by mail from the public information room. Include the project number, as well as your daytime telephone number, with your request. FERC will require at least several weeks to collect the materials, so you should leave plenty of lead time.

Using the Freedom of Information Act

Any written request for information should be filed under the Freedom of Information Act (FOIA),[1] which governs public access to federal documents. (The Commission's formal regulations explaining its FOIA procedures are included in Appendix E).[2]

A formal FOIA request is not legally necessary to obtain Commission documents that are kept in the public information room. You can obtain these documents by contacting the public information room, in person or in writing, and providing a reasonable description of the materials you seek. Nonetheless, it is good practice to invoke FOIA in any written request for documents. FOIA obliges the Commission to respond to your request within ten working days of receipt; it may entitle you to a waiver of copying costs; and you can appeal a denial of a request.

With few exceptions, the documents necessary for intervention are maintained in the Commission's public files. Types of documents available in the public information room are listed in the Commission's FOIA regulations.[3] These include project applications, motions to intervene, and other agencies' comments.

To obtain records not included in the public files, you must make a formal FOIA request. Reports on FERC's inspections of operating hydroelectric projects, for example, are generally not available in the public information room. Inspection reports can be particularly useful for developing your case in a relicensing proceeding if they disclose chronic safety problems with a project.

A FOIA request must be in writing, addressed to the Commission's director of public affairs, and prominently marked with the words *freedom of information request.* A FOIA request also must include either a statement that you are willing to pay a reasonable fee to obtain the materials, or a request for a waiver or reduction of the fees. (The fee can run into hundreds of dollars if the records are voluminous.) The Commission will reduce or completely waive the fee if you demonstrate that the information sought is (1) "in the public interest because it is likely to contribute significantly to public understanding of the operations or activities of the government" and (2) "not primarily in the commercial interest of the requester."[4] You have strong grounds for a fee waiver if you don't have a commercial purpose in seeking the materials and if the information will be used to educate the public, for example in your organization's newsletter.

If a FOIA request for documents is denied, or if the Commission staff denies a request for a fee waiver or reduction, you have the right to appeal the decision to the general counsel of the Commission. The forms and procedures for appealing a FOIA decision are set forth in FERC's regulations. If the denial of a request for records or a fee waiver or reduction is upheld in whole or in part on appeal, the general counsel is required to notify you of the opportunity to seek judicial review of the decision. You then have the option to pursue the matter in court.

INTERVENING IN A FERC PROCEEDING

To make sure that your comments about a project application will carry weight with the Commission, you must become an intervenor in the proceeding and support your argument with hard facts and sound logic. As an intervenor, you will have the following rights:

- You will receive all documents that the applicant and other intervenors file in the proceeding.
- You will receive the Commission's final decision on the application. Prior to that, you will be informed of any other major developments in the proceeding.
- You will have the right to seek judicial review of the Commission's final decision.

If you do not intervene, you may file comments with the Commission, but there is no guarantee that your comments will be taken seriously. Most important, if you do not intervene, you may not have the right to seek judicial review of the Commission's decision; the result is less leverage in the Commission proceeding itself and fewer options if you disagree with FERC's final decision.

Motions and Timing

You may file a *motion to intervene* any time after the application for a preliminary permit, license (original or renewed), or exemption has been filed with the Commission. Thus, you do not need to wait to file until the Commission has accepted the application for filing or has published the public notice. The advantage of intervening as early as possible is that you will receive all subsequent filings in the proceeding. On the other hand, you might overlook some issue that you would have raised in your intervention if you had waited to file until you were more educated about the project; the Commission may elect not to consider arguments not presented in the original motion to intervene.

The deadline for filing a motion to intervene is generally the *comment date* that appears in the public notice published in the *Federal Register* or local newspapers. (The notice usually sets the comment date between 45 and 60 days later.) Whenever possible, file your motion to intervene on or before this deadline.

If filed on time, your motion will be granted automatically if no party files an answer in opposition.[5] You will receive no notice from the Commission that your motion has been granted, but you will be a party to the proceeding. The Commission must affirmatively grant a motion to intervene if another party opposes your motion or if your motion is not filed on time.

There is a special, later deadline for filing a motion to intervene in the relatively rare proceeding in which an environmental impact statement (EIS) is prepared.[6] Any person who would otherwise be entitled to intervene in a proceeding may intervene on the basis of the draft EIS. So long as the motion to intervene is filed within the comment period for the draft EIS, the Commission considers the motion to be filed on time. Intervention at the EIS stage makes you a full party, entitled to participate in all subsequent phases of the proceeding.

Intervention in a preliminary permit proceeding does not constitute intervention in the subsequent licensing proceeding. Even if you have already intervened at the preliminary permit stage, you must file a separate motion to intervene in a licensing proceeding.

If you learn about a filing deadline before it has passed but cannot submit your intervention motion on time, you can file a motion for extension of time. The motion must specify how much additional time you need and the reasons for the delay: for example, that you need additional time to collect relevant facts. FERC may grant the motion if you show "good cause."[7]

If you have missed the deadline for filing an intervention motion you may still file a motion for permission to file a late intervention. Factors supporting late intervention include good cause for filing late; no disruption of the ongoing proceeding or prejudice to existing parties; and inability of the present parties to represent your interests. FERC may establish limits on late intervention, to avoid delay or prejudice to the existing parties.[8] While it is difficult to predict how the Commission will act on a particular motion for late intervention, it is known that FERC grants such motions infrequently. (But see The Story of Bulls Bridge Gorge on page 23.)

What the Intervention Should Contain

Appendix C is a model intervention motion. If you follow the form of that model and pay attention to the specific guidelines set forth here, you should have no difficulty obtaining intervenor status, so long as your intervention is filed on time. A motion to intervene intended simply to get your foot in the door can be much more succinct than the model. On the other hand, some motions to intervene run into hundreds of pages.

You do not need a lawyer to intervene. Preparing an intervention can be as simple as filling out the IRS short form. However, if you can find a lawyer (especially one with prior experience before FERC) to assist you in preparing the intervention, you should do so. Almost every hydroelectric proposal raises some unique legal issues that a lawyer is better equipped to handle than a layperson.

The following are the basic requirements for a motion to intervene:

1. *Format.* Your motion should start with a caption that states "United States of America, Federal Energy Regulatory Commission," followed by the project name and number. Next, you should include a heading that describes the filing (e.g., "Motion to Intervene of . . .") and that states your name or the name of your organization.[9] You may file a motion to intervene for yourself alone, for an organization, or for a coalition of any number of individuals and groups that have agreed to participate. Use 8½ by 11 inch paper stapled on the left-hand side.[10]
2. *Statement of interest.* The motion to intervene must contain sufficient factual detail to demonstrate that (1) the intervenor "has or represents an interest which may be directly affected by the proceeding" or (2) the intervenor's "participation is in the public interest."[11] You should argue if you have a basis for saying so, that you will be directly affected by the outcome of the proceeding and that your participation will serve the public interest. You will be directly affected, for example, if you use a river for boating, fishing, hiking, or virtually any other purpose and your enjoyment of the river will be affected by the proposed project. Your participation will serve the public interest if you have some particular knowledge or experience that will assist the Commission in resolving the factual or legal issues in the proceeding.
3. *Statement of position.* The motion to intervene must state "the position" of the intervenor and "the basis in law and fact for that position."[12] You should clearly state, for example, that you oppose issuance of the

preliminary permit, license, or exemption or that you do not oppose approval of the application on the condition that certain terms and conditions be included in the preliminary permit, license, or exemption.

4. *Signature.* Every motion to intervene must be signed. If you are filing a motion on behalf of a group, you should clearly indicate that you are signing the motion for the group.[13]
5. *Certificate of service.* Every motion to intervene must include a *certificate of service,* which testifies to the fact that you have served the motion, that is, you have mailed the motion to the applicant and any other parties to the proceeding, as required by the Commission's regulations.[14] The certificate must contain specific wording, which is included in the model intervention in Appendix C.
6. *Service.* You must send a copy of your motion to intervene (and of every other document you subsequently file in the proceeding) by first-class mail to the applicant and to the other intervenors in the proceeding no later than the date of filing.[15] The list of intervenors who are entitled to *service* in each proceeding is maintained by the secretary of the Commission. You can obtain a copy of the official service list from FERC's public information room. (Include the project number and your daytime telephone number with your request.) It is prudent to request the service list at least several days in advance if you will be picking it up in person or several weeks in advance if you are requesting it by mail.
7. *Filing.* You must file with the Commission the original and 14 copies of your motion to intervene (and of every other document you subsequently file in the proceeding).[16] If you wish to have a record of the Commission's receipt of your filing, you should supply a 15th copy and a stamped, self-addressed envelope. The Commission staff will mark the copy with the time and date and return it to you.

Getting Your Motion to FERC

To get your motion to intervene properly before the Commission, you must *file* it. You can file a motion to intervene in person by delivering the original and 14 copies to FERC's public information room, or you can file your motion by mail. If time is short, you can send the original and 14 copies by an overnight delivery service, to make sure they arrive at the Commission in time.

Note that a document is "filed" on the day it is received by the Commission, not the day it is mailed to the Commission. By contrast, service of a document is effected on other parties when it is deposited in the mail.

Next Steps

Once you have filed a proper motion to intervene, you are not required to take any further steps in the proceeding. The Commission must consider the comments and evidence in your intervention, along with those of the resource agencies and other intervenors, in acting on the application. If, after the Commission makes a decision, you are dissatisfied with the outcome, as a formal party to the proceeding you can appeal a staff decision, seek rehearing of a Commission order, or seek to overturn a Commission decision in court. The likelihood that your arguments will prevail before the Commission, or subsequently in court, depends on the strength of your legal logic and the quality of your factual evidence.

There are some steps you can take between the filing of your motion to intervene and the issuance of a Commission decision. If the Commission orders that an adjudicatory hearing be held, you are entitled to participate in the hearing. Furthermore, if the Commission prepares an environmental impact statement on the project, you should review the draft EIS and submit comments to the Commission. You may participate in the hearing that the Commission holds on the draft EIS.

Also, you may submit additional materials to the Commission to support arguments you have presented in your motion to intervene. For example, if you did not have an opportunity to conduct a survey of white-water boaters prior to filing your motion to intervene, you can do so after the filing and submit the results in the form of a report to the Commission. It is important that you submit such information to the Commission before the staff completes its own environmental analysis.

You can file a motion with the Commission requesting that FERC take a particular action. For example, if you have suggestions as to how the Commission should analyze a project's cumulative environmental impacts, you can file a motion for a *cumulative impacts analysis.* If you conform to the Commission's technical requirements of filing and service, you can allow your common sense and imagination to guide you on how best to proceed.

Your additional filing need not explain in detail your interest in the proceeding. You only need to observe that you are a party to the proceeding. However, you must be sure to comply with the other technical rules for FERC filings, as already described.

As a general rule, no party to a proceeding may directly contact a FERC Commissioner or staff employee to discuss the review of a project unless the communication is "on the record"—for example, a motion filed with the Commission and served on all the parties.[17] This prohibition against *ex parte* contacts is intended to prevent secret influence on FERC's deliberations.

You can communicate with FERC employees by telephone or in person in several limited circumstances: a request for advice or assistance unconnected to any ongoing regulatory proceeding; an inquiry about the status of a proceeding or about a procedural matter (for example, the deadline for a responsive pleading, the date of a hearing, or the expected release date of an EIS); any communication covered by an agreement between all parties that any party may contact FERC on certain matters without notice to the other parties; or a request for supplemental information or data necessary for an understanding of documents already in the record, provided the request is made in the presence of a FERC staff attorney assigned to the proceeding or after coordination with the attorney.[18]

FIGHTING FOR THE AMERICAN

Harriet LaFlamme and Dawn King are two California school teachers who, until recently, had never heard of FERC. But then they discovered that Joseph Keating wanted to divert a mile of their favorite river for the Sayles hydro project. The river was the South Fork of the American River in the Sierra east of Sacramento. They got to work learning the FERC process, and before too long, Harriet and Dawn won a precedent-setting victory against FERC in court. With a little more work and good luck, they just might save the South Fork.

Harriet and Dawn first learned about the project at a public meeting, held by the U.S. Forest Service in the El Dorado National Forest, regarding another proposed hydro project on Pyramid Creek. The message from the Forest Service was that the Pyramid project was supported by all government agencies, and it was going to be built. (The Sayles project on the South Fork of the American River was kept hush-hush, not even discussed, but a few of the attendees had heard about it through an Army Corps of Engineers notice that went to cabin owners.) Harriet and Dawn refused to believe that the projects were inevitable. And they spurned a number of suggestions over the next few years that they were simply wasting their time.

The first step was to obtain intervenor status. Even though they had missed the formal deadline—in fact, neither Harriet nor Dawn had seen the small-print legal notices in the local mountain newspapers—they filed for leave to intervene anyway: Harriet on Sayles, Dawn on Pyramid. Even though they were technically late, the Commission granted their motions to intervene. About six months later, FERC issued the license on Sayles. In response to Harriet and Dawn's inquiries, a helpful FERC staff attorney coached Harriet through the process of preparing and filing a petition for rehearing, literally drawing a picture of the format, with margin widths, and so on.

Almost two years later, FERC denied Harriet's petition for rehearing. Enter, through the help of the California Save Our Streams organization, Glen Kottcamp, a dedicated environmental lawyer with substantial expertise in FERC issues, who brought Harriet's case to the U.S. Court of Appeals for the Ninth Circuit.

In March 1988, the court granted Harriet a stunning victory, overturning the license order and sending it back to the Commission [*LaFlamme v. FERC*, 842 F.2d 1063 (9th Cir. 1988)]. Basically, the court ruled that FERC had violated both the National Environmental Policy Act and the Federal Power Act by failing to prepare an environmental impact statement that looked at the project's serious effects on the scenic and recreational resources of the area (see chapter 3). In addition, the court invalidated the license on the grounds that FERC had not prepared a comprehensive plan that addressed the relationship between the Sayles project and other development projects in the entire river basin (see chapter 6). The court also ruled that FERC's policy of licensing first and studying later was illegal.

While the appeal was pending, Joseph proceeded full tilt with the construction of the Sayles project. In the absence of a stay of the license order from the Commission or a court, he was legally entitled to proceed. By the time of the court decision, the project was 90 percent complete.

FERC was put under a court order to conduct a proper environmental review and to reach a proper decision on the project. The Commission announced that it would be considering two basic alternatives—completion of the dam with additional mitigation measures, and complete removal. At a public "scoping session" in Placerville, California, in October 1988, scores of citizens turned out in almost unanimous support of dam removal. In May 1989, the FERC staff released an environmental assessment recommending that the project be completed.

RECTIFYING A FERC ERROR

If the Commission grants a license despite your arguments against it, you have an avenue of redress: you can appeal a decision by the Commission staff, seek a rehearing of an order from the full Commission if your appeal is denied, and finally, attempt to overturn a Commission decision in court.

Appealing a Decision by the Director

If the director of the Office of Hydropower Licensing issues an order with which you disagree, you can file a *petition for appeal* with the full Commission. Also, if you intend to seek judicial review of a staff order—or even if you merely want to keep that option open—you must file a petition for appeal with the Commission. In the absence of a petition for appeal, a staff order becomes the final decision of the Commission. You do not need to be an attorney to file a petition for appeal.

You may wonder, when you receive a particular order, whether it is a staff order or a Commission order. You can identify a staff order by the signature of the director of the Office of Hydropower Licensing. A Commission order will be signed by the secretary of the Commission.

A petition for appeal must be filed within 30 days of the issuance of the staff order.[19] Any party to the proceeding may file an answer to an appeal within ten days of service of the appeal.[20] In accordance with the general rules for filings, you must file an original and 14 copies of the appeal, with a certificate of service, and then serve the appeal on the applicant and any other parties on the service list, and so on.

Pete Hall

The Sayles Flat Project on the American River in California. A court decision overturning the Commission license halted construction.

If the Commission takes no action within 30 days after you file the appeal, the appeal is deemed denied.[21] Thus, the Commission must grant the appeal or take some other affirmative action to prevent an automatic denial. The Commission commonly grants an appeal solely for the purpose of extending the 30-day period and allowing itself more time to consider the merits of the appeal.

The Commission's denial of a petition for appeal (either by formal Commission action or by the running of the 30-day period) is considered to be a final decision that is subject to a request for rehearing.

Appealing a Decision of the Full Commission

After the Commission has issued an order in a hydroelectric proceeding, including a decision on an appeal of a staff action, any party may *petition for rehearing* of the order by the Commission.[22] The filing of a petition for rehearing is an essential prerequisite for later seeking judicial review of the Commission's order. If you fail to file a petition for rehearing, your right to appeal may be foreclosed.

A petition for rehearing must be filed "within thirty days after the issuance of such order."[23] The petition must state the grounds for seeking rehearing "concisely"[24] and must "set forth the matters relied upon by the party requesting rehearing, if rehearing is sought based on matters not available for consideration by the Commission at the time of the final decision or final order."[25]

The Commission "will not permit" the other parties to answer a petition for rehearing.[26] At its discretion, the Commission "may afford parties an opportunity to file briefs or present oral argument on one or more issues presented by a request for rehearing."[27] The Commission rarely exercises this power and usually disposes of a petition in a simple written order. The Commission can, and often does, reject a petition in part and grant it in part, modifying the original Commission order accordingly.

While a Commission order can be challenged immediately through a petition for rehearing, you must challenge a staff order by first filing a petition for appeal, as already discussed; only if the Commission denies the appeal can you file a petition for rehearing. That is, a party may not seek rehearing of a staff action unless that party or another party to the proceeding has appealed the staff action to the Commission.

A petition for rehearing is deemed denied unless the Commission acts upon the request within 30 days after the request is filed.[28] The Commission routinely grants so-called tolling orders, which extend the

30-day period until the Commission has had an opportunity to consider the request for rehearing on its merits.

Obtaining a Stay of an Order

A request for rehearing is of limited benefit if the Commission order remains in effect and construction of the project you oppose can proceed while the Commission ponders your petition. (See Fighting for the American on page 88.) To preserve the status quo, a party may apply to the Commission for a *stay* of the effective date of the Commission order, pending resolution of the request for rehearing.[29]

The Commission grants a stay when "justice so requires."[30] This broad standard, derived from the Administrative Procedure Act, means that the Commission "considers such things as: whether the movant will suffer irreparable injury in the absence of a stay; whether the issuance of a stay would substantially harm other parties; and where the public interest lies."[31]

VIGILANCE AFTER PROJECT CONSTRUCTION

It is not uncommon for a developer to propose a significant change in project design or operation after the license or exemption has been issued. You should be alert to this possibility, to ensure that whatever success you have achieved in the regulatory process will not be undone. An intervention in a licensing proceeding expires once the license is issued, so you must intervene again in any proceeding to amend the license or exemption.[32]

The Commission has historically approved projects without fully analyzing their environmental impacts. Instead of requiring specific mitigation of environmental damage, the license or exemption directs the applicant to perform studies for the purpose of formulating terms and conditions. The obvious flaw in this practice, which the courts have condemned,[33] is that a misguided approval cannot be reversed once a project has been constructed.

There are several things you can do to retain some influence in the postlicensing stage. First, in your motion to intervene in the licensing proceeding, you can request that the Commission include a condition that requires the project owner to notify all intervenors of any proposed amendments filed with the Commission. In the absence of such a condition, the owner would be obliged to inform only the Commission of the proposed amendment, regardless of the potential to affect the interests of the original intervenors.

Second, also in your motion to intervene, you can request that you or your group be specifically named as one of the entities (along with appropriate federal and state resource agencies) that the project owner must consult if the Commission orders postlicensing studies on issues of particular interest to you. A typical condition on a license requires the project owner to study a particular impact, develop mitigating measures, consult various agencies and entities about the study and the proposed mitigation, and then submit the results to the Commission. If your group is concerned about white-water boating, for example, you can ask that your group be named as one of the entities to be consulted if the Commission orders postlicensing studies related to such recreation.

Finally, if you believe a developer is violating the terms of a license, you can telephone the Commission's Division of Project Compliance and Administration in the Office of Hydropower Licensing. You may also file a written complaint with the Commission.[34] The basic rules for preparing and filing a complaint are the same as those for motion to intervene. The developer must file an answer to the complaint, unless the Commission orders otherwise,[35] and the Commission may initiate a formal factual investigation to determine the merits of the complaint.

REFERENCES

1. 5 U.S.C. § 552.
2. 18 C.F.R. §§ 388.101–388.112.
3. 18 C.F.R. § 388.106.
4. 18 C.F.R. § 388.109(b)(6).
5. 18 C.F.R. § 385.214(c).
6. 18 C.F.R. § 380.10.
7. 18 C.F.R. § 385.2008(a).
8. 18 C.F.R. § 385.214.
9. 18 C.F.R. § 385.2002.
10. 18 C.F.R. § 385.2003.
11. 18 C.F.R. § 385.214(b)(2)(ii),(iii).
12. 18 C.F.R. § 385.214(b)(2).
13. 18 C.F.R. § 385.2005.
14. 18 C.F.R. § 2010.
15. 18 C.F.R. § 2010.
16. 18 C.F.R. § 385.2004.
17. 18 C.F.R. § 385.2201.
18. 18 C.F.R. § 2201(b).
19. 18 C.F.R. § 385.1902(b).
20. 18 C.F.R. § 385.1902(b).

21. 18 C.F.R. § 385.1902(c).
22. 16 U.S.C. § 825(1) and 18 C.F.R. § 385.713.
23. 16 U.S.C. § 825(1).
24. 16 U.S.C. § 825(1) and 18 C.F.R. § 385.713(c)(1).
25. 18 C.F.R. § 385.713(c)(3).
26. 16 U.S.C. § 385.713(d)(1).
27. 16 U.S.C. § 385.712(d)(2).
28. 18 C.F.R. § 385.713(f).
29. 18 C.F.R. § 385.713.
30. *In re City of Fort Smith,* 43 FERC ¶ 61,238 (CCH) (11 May 1988).
31. 43 FERC at 61,646.
32. See *In re Kings River Conservation District,* 36 FERC ¶ 61,365 (CCH) (29 Sep. 1986) (explaining Commission standards for when proposed amendments require that public be given new opportunity to intervene).
33. See *Confederated Tribes and Bands of the Yakima Indian Nation v. FERC,* 746 F.2d 466 (9th Cir. 1984).
34. 16 U.S.C. § 825(e).
35. 18 C.F.R. § 385.206(b).

Chapter 6

Raising Issues before FERC

Supporting your arguments with facts • Focusing on procedure as well as substance • Encouraging allies to make similar arguments • Citing legal authorities • Methods for dispute resolution • Comprehensive river basin plans • Impacts on natural resources • Environmental impact statements • Consistency with federal land management plans • Energy generation, project cost, and finances

INTERVENOR STATUS ONLY GETS you in the door. To get the Commission to accept your point of view, you must use the intervention motion and any later filings to present your best arguments and evidence regarding the merits of the applicant's proposal.

Each project on each river is different. Accordingly, you must develop your arguments in response to the project itself and the specific impacts that concern you. Since the standard of "public interest" under the Federal Power Act is all-encompassing, the Commission must consider any argument that, as a matter of common sense, bears on whether, or on what conditions, a project should be approved.

Thus, for example, if the key issue is the project's potential impact on archeological relics, you should emphasize that. You should focus on a project's economic impacts if it would eliminate a stretch of white water that supports local rafting companies.

Certain issues, however, have arisen repeatedly in hydroelectric proceedings. These have proven important to the Commission in its evaluation (and, on occasion, rejection) of project applications and in subsequent judicial review of Commission decisions. They are discussed here in *Potential Issues to Raise.* You should consider whether these precedents may apply to your proceeding.

GENERAL RULES FOR MAKING YOUR CASE

Support Your Argument with As Many Facts As Possible

The Commission's most common justification for dismissing conservationists' objections to a hydroelectric proposal is that the intervenors have failed to present factual evidence to support a particular argument. The Commission routinely refuses to deal with the merits of an argument on the ground that the argument is "mere opinion" or an "unsubstantiated allegation."

Consistent with its view of itself as a quasi court, the Commission believes it has little, if any, obligation to research and collect data necessary to resolve concerns raised by intervenors. The Commission expects the public (and other federal and state agencies) to do much of the work necessary to identify and attempt to resolve such concerns. Unless your intervention presents factual evidence and is based on legal arguments, the Commission is likely to disregard your position.

Accordingly, support your arguments with facts that are material to the proceeding. Appropriate evidence varies, depending on the issue. To demonstrate the value of a river for fishing, you might conduct a survey (preferably with some expert assistance) to estimate the numbers of anglers using the river and their expenditures in local communities, to assess the recreational merits of that river versus others in the region, and to gather the anglers' opinions about the proposed development. To demonstrate the scarcity of adventure white water or spectacular waterfalls, you could present data on the miles of white water or number of waterfalls in the region of the proposed project.

Focus on Procedural As Well As Substantive Issues

The Commission is concerned about form as well as substance. The Federal Power Act and the Commission's regulations are a maze of procedural requirements that project applicants and intervenors alike must satisfy. These requirements create pitfalls for intervenors and project applicants.

If you believe the Commission or an applicant has violated a procedural rule, use that to your advantage. Alert the Commission to any technical violation that affects the fairness of the regulatory proceeding or your participation. Following your intervention, you can submit a motion to rescind the acceptance of an application as properly filed, for example, if you believe that the exhibits do not contain all of the types of information required by the Commission's regulations.

Encourage Your Allies to Make Similar Arguments

There is strength in numbers. The Commission pays greater attention to an argument made by many groups and individuals. If you know that other conservationists plan to file intervention motions, you should coordinate the drafting so that you take consistent positions.

Discuss your intervention with federal and state agencies whose jurisdiction includes the project location or impacts. Such agencies routinely comment on—and sometimes intervene in—hydroelectric proceedings. At the national level, these include the U.S. Fish and Wildlife Service, the National Marine Fisheries Service (if the project affects anadromous fish), the U.S. Forest Service, the Bureau of Land Management, the National Park Service, and the U.S. Environmental Protection Agency. At the state level, the state fish and game agency, the pollution control agency, the parks department, and others may comment or intervene. You should contact the staffs drafting agency comments and discuss the merits of the project application and their agencies' positions.

In turn, you can discuss in your intervention the impacts on resources under the jurisdictions of federal and state agencies. Private individuals or groups sometimes argue the agencies' cases more forcefully than the agencies themselves, given political and institutional constraints.

Cite Legal Authorities

FERC follows the law as it interprets it. Cite statutory and regulatory provisions, court decisions, and FERC's own decisions on other applications, to justify your own interpretation of what the law requires for the disposition of the pending project application.

If you are not a lawyer, you may be surprised by the conflicts between the laws (as an environmentalist or like-minded court would interpret them) and FERC's rulings. But conflict is the nature of the law. Different courts rule differently on the same issue, and the Commissioners now constituting FERC may revise or ignore policies established by past Commissioners. This is not to say that you are participating in a random or arbitrary system of law. Your job is to persuade the Commissioners that your interpretation of the law, or its application in a particular proceeding, is better than some other.

Be accurate in your quotations of legal authority. Be fair in your explanation of what an authority means. If you aren't a lawyer, don't worry too much about the format for citations; try to follow the models in this book's footnotes and in the model intervention (Appendix C).

If you doubt that you can stop a project, or if you support a project on certain conditions, then include in your intervention (or a subsequent filing) recommendations for conditions to minimize adverse impacts on the environment or community. FERC includes standard conditions in licenses and exemptions;[1] use those as a starting point, and identify any special conditions that you believe are appropriate. In one instance, FERC required a developer to create alpine meadows in direct proportion to such habitat flooded by a storage facility, and to pay the local county for services during and after construction, including fire and police protection, roads, and school expansion for the construction workers' children.[2]

POTENTIAL ISSUES TO RAISE

Raise specific issues of fact and law. This section outlines the legal issues which you might ask FERC to address, depending upon the project. It is like a restaurant menu. Use what is appropriate in a specific proceeding; save the remainder for another day.

Methods for Dispute Resolution

You should recommend that FERC hold an *evidentiary hearing* to resolve disputed issues of fact. You should also oppose FERC's historical practice of deferring the establishment of project-specific terms and conditions until after licensing if you believe that the choice of such mitigation is relevant to the Commission's decision whether to even approve the project application.

Evidentiary Hearings

Under the federal Administrative Procedure Act, a Commission order disposing of a hydropower application must be "determined on the record after opportunity for an agency hearing."[3] This means that the Commission must conduct an *adjudication,* or evidentiary hearing, to resolve any disputed issues of material fact.

Your participation in a hearing, while costly and time-consuming, provides an important opportunity to present your side of the case. Also, the Commission's decision to conduct an evidentiary hearing can be a signal that the Commission has serious reservations about a project. Further, delay can be beneficial if your goal is to stop the project; an adjudicatory hearing can buy you months or years.

The Commission will schedule an evidentiary hearing only if it concludes that there is a concrete factual dispute, not to explore legal or policy issues.[4] Furthermore, the Commission will not schedule a hearing unless an intervenor has shown that the disputed issues of fact are *material:* that is, addressing them would assist the Commission in determining whether to grant the project application.[5]

To avoid the accusation that you have slept on your rights, include in your motion to intervene at least a brief request for an evidentiary hearing and an explanation as to why a hearing is necessary. If a hearing later proves unnecessary, you can choose not to press the request. The failure of the Commission to hold a hearing, when the factual issues have not been resolved, may provide an effective basis for seeking judicial review of a Commission order to which you object.

Deferral of Material Issues until After Licensing

The Commission has historically approved licensing applications before determining what mitigation of environmental impacts is appropriate. It relies on postlicensing studies to develop mitigation measures. This practice has resulted in the construction of projects for which mitigation of certain impacts proved impossible or prohibitively expensive. Such approval of projects, given incomplete knowledge of feasible mitigation, seems inconsistent with the Commission's obligations to determine whether a project as a whole is in the public interest.

In *Confederated Tribes and Bands of the Yakima Indian Nation v. FERC*,[6] the court reversed a Commission order granting a license, in part on the ground that the Commission had failed to address the project's adverse impacts on anadromous fish prior to licensing and had deferred consideration and implementation of fish protection measures until after licensing. The court held that, in order to carry out its public interest and comprehensive planning responsibilities, the Commission "must consider fishery issues *before*, not after, issuance of a license." The court's reasoning applies not only to fish protection, but also to management of other natural resources, including white-water boating, scenery, biological diversity, and historic and cultural artifacts.

You should consider whether the Commission has deferred resolution of environmental concerns when it issues an order on your project. Violation of the holding in Yakima may constitute grounds for administrative, and possibly judicial, appeal.

Comprehensive River Basin Plans

The Commission's Own Plans

One of the Commission's most fundamental responsibilities in reviewing a hydroelectric proposal is to determine whether the proposed project "will be best adapted to a comprehensive plan."[7] Recent court decisions, including *LaFlamme v. FERC*,[8] have stressed that the Commission must conduct "coordinated study and comprehensive planning along an entire river system before projects are authorized." Such a plan should address all uses and values of the river, including energy, recreation, and water supply and quality. Unfortunately, the Commission has invested almost no effort in the preparation of comprehensive plans in hydropower proceedings.

If the project would affect other uses of the basin or if you believe that it would result in excessive development of the basin as a whole, use your intervention or a subsequent motion to request that FERC prepare a written, comprehensive plan for the basin. Since the Commission has resisted court orders directing it to prepare comprehensive plans, it may not grant your request. That, in turn, could give you a strong ground for appeal.

Consistency with Other Federal and State Agencies' River Plans

Under the Electric Consumers Protection Act (ECPA), the Commission must consider the extent to which a proposed project is consistent with a comprehensive river plan prepared by the state or by another federal agency. Especially if the agency that prepared the comprehensive plan is not actively participating in the proceeding, you should address the project application's compatibility with that plan.

Virtually any type of resource plan that relates to rivers qualifies as a comprehensive plan within the meaning of ECPA. However, the Commission recognizes a state or federal plan as a comprehensive plan under ECPA only if it is actually on file with the Commission. If you identify a comprehensive plan that could affect FERC's consideration of a project application but is not on file with the Commission, you can file it yourself, either as part of your intervention application or by subsequent motion.

Impacts on Natural Resources

In applying the *public interest* and *comprehensive planning* standards under the Federal Power Act, the Commission has the authority to veto a proposal for development that would result in unmitigatable damage to valuable natural resources. While this principle is often honored in the breach, the Commission is bound by its own precedents to reject a project application if intervenors demonstrate that the benefits of development are outweighed by the benefits of preserving a free-flowing river.

Before the 1980s, the Commission rejected only one hydropower proposal on environmental grounds, where it had legal discretion to approve or reject the application. This proposal involved the Namekagon River, a tributary of the St. Croix River, in Wisconsin. The threatened stretch of river was particularly beautiful, with gentle white water accessible to large numbers of boaters and a nationally recognized small-mouth bass fishery. These resources were becoming increasingly rare in Wisconsin: the evidence showed that Wisconsin already had lost 92 percent of its original heritage of free-flowing rivers to dams and other river development. The federal appeals court rejected the disappointed applicant's appeal and upheld the Commission's conclusion "that the unique recreational features of the river were of greater public benefit than the use of the river for water-power development."[9] The Commission reached this conclusion even though the applicant demonstrated the project's engineering feasibility, its ability to finance the project, and a regional need for the power the project would have generated.[10]

In 1987 the Commission applied a similar analysis in denying an application to construct a hydroelectric project on the Kootenai River at Kootenai Falls in Montana.[11] The Commission reasoned that the project would have destroyed the scenic beauty of the falls, adversely affected the rainbow trout fishery, and interfered with the religious practices of the Kootenai Indians.

The project would have reduced the scenic splendor of Kootenai Falls by diverting most of the water out of the river. The Commission relied on an on-site survey conducted by the state of Montana to document the large numbers of people who visit the falls. An inventory of other falls in surrounding states demonstrated that Kootenai Falls is one of the last undeveloped waterfalls on a major river in the northwestern United States.

The stretch of river threatened by the project had developed into one of the best rainbow trout fisheries in Montana, as a result of careful regulation of flows from an upstream dam. The project would have inundated four miles of riverfront and significantly reduced the river flow below the proposed dam. While the impact below the dam could have been lessened by the setting of a minimum downstream flow, the Commission concluded that such mitigation would have rendered the project uneconomic.

Finally, the project would have interfered with the Kootenai Indians' religious practice of "vision questing," in which they spend long periods in the wilderness seeking revelations. Vision questing is possible only at specific consecrated sites, and the proposed project would have occupied one of the most significant such areas, believed to be the home of an Indian deity. The Commission found the project to be incompatible with continued religious use of the area.

Four project applications in the Owens River basin in California were rejected in early 1988.[12] The Commission identified the likely adverse impacts of these and several other projects and determined what types of mitigation measures (stream flow requirements, burial of project facilities to minimize visual impact, and so on) would be necessary to eliminate the impacts. The Commission then assessed whether the projects would remain economically viable if the necessary mitigation measures were applied.

The Commission denied the application for one project because the

Commission found that (1) certain minimum flows would have to be released below the project to mitigate impacts to streamside vegetation and (2) the project would have to pump water to an irrigation system that would otherwise be adversely affected by the project. These mitigation measures together rendered the project infeasible.

As to the other three projects, the Commission determined that one or more of the following measures would be required: minimum flows to maintain wild trout populations, burial of the project penstocks, and repositioning of the project structures either up or down the river. The Commission concluded that these mitigation measures would render these projects uneconomic and therefore rejected these applications as well.

OPEN SEASON ON WHITE WATER

It's 8:00 A.M. The dam keeper at the reservoir turns a big wheel, opening the valves. Frigid water suddenly surges from the bottom of the lake, 1,000 cubic feet per second rushing downstream. Within seconds, a quiet little stream becomes a raging torrent. Water chills the humid summer air, making a dense cloud of fog over the river canyon.

An air of excitement and tension silences thousands of spectators lining the riverbank. All eyes focus on a kayaker floating nervously in a river pool. He wears a white vest, number 1. Suddenly a gun goes off. The kayak charges downriver at breakneck speed, disappearing into the churning foam.

The event is the World Canoe and Kayak Championships, held in 1989 on Maryland's Savage River, one of the best white-water race courses in the world.

Incredible as it may seem, while plans for the race were being made, white-water aficionados discovered that several hydropower developers were eyeing the Savage. Preliminary permits, and even an application for an exemption, had been filed. The developers planned to be in construction right in the middle of the scheduled 1989 world championships! If the project were built, the conditions necessary for power generation would be totally incompatible with the high flows needed for racing.

The Savage is not a federal wild and scenic river, nor is it in a national park or wilderness area. This meant one thing to the developers: open season!

Sensing trouble, white-water groups sought advice. Expert Washington hydropower lawyers Ron Wilson and Chuck McGraw said they could slow down the project, and possibly derail it, but it would cost money.

Abby Endicott

The Savage River in Maryland. The site of the 1989 World Canoe and Kayak Championships is threatened by a hydropower proposal.

While the lawyers went to work, Mac Thornton, an ebullient workaholic kayaker from the D.C. area, put together a volunteer fund-raising group. They decided on a raffle because it would simultaneously raise money and publicize the problem. Thornton recruited an army of volunteers. They staffed phone banks for months on end, calling every canoeist and kayaker in the region. Thornton staged a media blitz so that when the raffle prizes were finally awarded, newspaper reporters were on hand.

Meanwhile, FERC denied the preliminary permit but kept the exemption application under consideration. Then it was learned that the developer would need state lands for the powerhouse. Normally, developers of FERC projects can confiscate any land they need from the owner—even a state—if they get approval from FERC. But there was one hitch. This confiscation power applies only to licenses—not exemptions. The state refused to give up its land, and FERC had to throw out the exemption application.

A happy ending? Not quite. Within days of throwing out the exemption, FERC reinstated another developer's old preliminary permit application. When this book went to press, exhausted white-water enthusiasts were starting all over again.

Environmental Impact Statements

Your motion to intervene should clearly state your recommendation on the appropriate kind of environmental analysis required by the National Environmental Policy Act (NEPA). If you think an environmental impact statement (EIS) is necessary, you should say so and explain in as much detail as possible why you think the project would significantly affect the environment. As discussed in chapter 3, an EIS must be prepared if FERC believes that a project could significantly affect the environment; if FERC expects a project to have a lesser impact, then it prepares a less detailed document, called an environmental assessment; and if it appears that a project would have an insignificant impact, then FERC's obligations under NEPA can be satisfied without any further environmental review.

On several occasions, courts have overturned FERC orders which violated NEPA. In *The Steamboaters v. FERC,*[13] the court ruled that the Commission had not adequately explained its decision not to prepare an EIS when it simply stated in the licensing order that, based on the terms and conditions included in the order, the project would not significantly affect the environment. In *Confederated Tribes and Bands of the Yakima Indian Nation v. FERC,*[14] the court rejected the Commission's argument that an EIS was not required in connection with a relicensing because the dam already had existed for 50 years. The court ruled that an EIS was necessary to address the significant opportunities for improving environmental conditions through relicensing.

Examine the Commission's NEPA regulations to determine what level of analysis would ordinarily be appropriate for the project of concern to you. (See Appendix E for a copy of the Commission's NEPA regulations.) At a minimum, you should insist that the Commission conduct the level of environmental analysis required by its own regulations.

The Commission picks the appropriate level of environmental analysis on the basis of the particular project's circumstances. The guidelines of the Council on Environmental Quality (CEQ),[15] which are binding on the Commission, state that this determination should take into account "context" and "intensity."[16]

Context means the setting of the proposed action; in other words, a project proposed for a blue-ribbon trout stream requires more careful analysis than a project proposed for a river without any outstanding characteristics. *Intensity* or severity of environmental impact refers to such factors as the uniqueness of the geographic area affected, the degree to which the proposed action is likely to be controversial, and the project's relation to other actions with cumulatively significant impacts.

Consistency with Federal Land Management Plans

The Commission may license a project within a federal "reservation" only if it can find "that the license will not interfere or be inconsistent with the purpose for which such reservation was created or acquired."[17]

On several occasions, the Commission has denied hydroelectric applications because of inconsistency with federal land management objectives. In the case of *In Re Rainsong Co.*,[18] the Commission found that a project proposed for Lena Creek in Washington would have deviated from the U.S. Forest Service's policy against road construction and logging in the Lena Lake special management area, part of the Olympic National Forest. Importantly, the Forest Service had commented that the project would be inconsistent with its management policy for that area.

Contact the local office of the federal land manager (usually the Forest Service or the Bureau of Land Management) if you believe a project would be inconsistent with the present character and management of a federal reservation. Try to enlist the federal agency's aid in intervening. The Commission makes the required consistency determination under the Federal Power Act, but its judgment is strongly influenced by the land manager's own conclusions about conflicts with land management objectives.

As discussed in chapter 7, the Forest Service and the Bureau of Land Management maintain that the Federal Land Policy and Management Act gives them independent authority over hydroelectric development, although the Commission disputes that authority. Furthermore, as discussed in chapter 3, the federal land management agencies have authority under the Federal Power Act to impose mandatory conditions on projects on federal lands.

Energy Generation, Project Costs, and Finances

Congress has given FERC wide discretion in judging the economic worth of a proposed hydroelectric project. The original Federal Power Act was based on the premise that "the utilization of . . . waterpower, which is inexhaustible and would not diminish by use, would add inestimably to the comforts of life and the promotion of public welfare. . . ."[19] Subsequent amendments have implicitly acknowledged the competing values of natural resources and have directed FERC to strike a balance in the public interest, consistent with comprehensive plans for the river basin.

Need for Power

The Public Utility Regulatory Policies Act (PURPA) guarantees a market for the electricity generated by a qualifying facility, one which has a capacity under 80 megawatts and which meets PURPA's and ECPA's tests for minimal environmental impact. The local utility must purchase this project's electricity at "avoided cost," namely, the cost of generating the same amount of electricity with alternative fuel. A nonutility applicant for a qualifying facility could thus receive approval for its application without any showing as to electricity demand.[20]

However, for any project that does not qualify for PURPA's subsidy, the applicant must demonstrate that the project would serve an actual or reasonably foreseeable demand for electricity. In the absence of such a need, the Commission must generally reject the proposal. For example, an application by the Rocky Mountain Power Company was rejected because "all of [the applicant's contract negotiations with utilities] are in the proposal state and Applicant has put forward no tangible evidence which would allow us to reasonably conclude that, based on its present efforts, a market exists or will exist for project power."[21] Examine the need for the power that would be generated by a proposed project not qualifying under PURPA, since many regions now have a substantial excess of generating capacity.

In *Idaho Power Co. v. FERC*,[22] the U.S. Court of Appeals for the Ninth Circuit upheld the Commission's rejection of Idaho Power Company's application to construct a project on the Snake River. The court agreed the applicant had not demonstrated demand for the project's electricity. The court ruled that the Commission has a responsibility to examine the need for power in the region of the proposed project: the Federal Power Act directs the Commission to license a project only if "necessary or convenient,"[23] that is, if it would serve the public interest. The court concluded that the Commission properly found that the Snake River project was not in the public interest because Idaho Power's forecasts showed no need for new electricity supplies until many years after the project would have been completed.

Project Cost and Feasibility

Is a proposed project economically viable, that is, is it a workable investment from the financial standpoint of the applicant? This question is important to conservationists because a bankrupt or hard-pressed developer may not properly operate the project or pay for essential environmental mitigation. The burden of proof to establish financial feasibility is on the developer seeking the license or exemption.[24] Do not assume that the applicant has protected its own self-interest by accurately calculating the costs and revenues of the project.

The Commission has rejected project applications where the applicant has not demonstrated its ability to obtain necessary financing[25] or included current estimates of a project's cost[26] or demonstrated that the annualized cost of generating electricity is less than the local utility's alternative cost of generation.[27]

If the applicant uses a benefit-cost analysis, which compares a project's expected benefits and costs, review it to ensure that *external costs*—e.g., those losses imposed on society—are taken into account. Resource economists have developed several techniques for quantifying the values of natural resources.[28] As discussed in chapter 8, you may need expert assistance to make a convincing case regarding project economics.[29]

REFERENCES

1. See 18 C.F.R. § 2.9, which provides citations to various FERC reports for the standard licensing conditions. See 18 C.F.R. § 4.94 for a conduit exemption's standard conditions; 18 C.F.R. § 4.106 for a case-specific exemption's standard conditions.
2. *In re Calveras County Water District,* 18 FERC ¶ 61,124 (CCH) (8 Feb. 1982).
3. 5 U.S.C. § 554(a).
4. *In re Tennessee Gas Pipeline Co.,* 26 FERC ¶ 61,144 (CCH) (6 Feb. 1984).
5. *In re Escondido Mutual Water Co.,* 20 FERC ¶ 61,157 (CCH) (5 Aug. 1982).
6. 746 F.2d 466 (9th Cir. 1984).
7. 16 U.S.C. § 803(a)(1).
8. 842 F.2d 1063 (9th Cir. 1988).
9. *Namekagon v. Federal Power Commission,* 216 F.2d 509, 512 (7th Cir. 1954).
10. *In re Namekagon Hydro Co.,* 12 FPC 203 (30 July 1953).
11. *In re Northern Lights,* 39 FERC ¶ 61,352 (CCH) (25 June 1987).
12. *In re Joseph M. Keating,* 42 FERC ¶ 61,030 (CCH) (20 Jan. 1988).
13. 759 F.2d 1382 (9th Cir. 1985).
14. 746 F.2d 466 (9th Cir. 1984).
15. 40 C.F.R. § 1500.1.
16. 40 C.F.R. § 1508.26.
17. 16 U.S.C. § 797(e).
18. 41 FERC ¶ 62,442 (CCH) (29 June 1987).
19. H. R. Rep. No. 66, 66th Cong. 1st Sess. (1916).
20. PURPA was intended "to encourage the development of such facilities, even in the absence of power deficits, in order to displace other sources of power." *Idaho Power Co. v. FERC,* 767 F.2d 1359, 1362 (9th Cir. 1985).
21. *In re Rocky Mountain Power Co.,* 37 FPC 900, 901 (15 May 1967).
22. 767 F.2d 1359 (9th Cir. 1985).
23. 16 U.S.C. § 797(e).
24. *In re Rocky Mountain Power Co.,* 37 FPC 329 (16 Feb. 1967).
25. *In re Robert P. Wilson,* 28 FPC 571 (10 Oct. 1962); *In re Public Power and Water Corp.,* 12 FPC 197 (15 July 1953).
26. *In re Calveras County Water District,* 51 FPC 895 (1 Mar. 1974).
27. *In re Public Utility District No. 1,* 39 FERC ¶ 62,057 (CCH) (10 Apr. 1987). The Commission calculated that the proposed project on the Lyre River in Washington would generate electricity at a cost of 8.16 cents per kilowatt-hour, more than the market value of 7.13 cents in the Northwest.

   In most cases, the Commission follows the simplistic approach of comparing a project's costs with the costs of power available from other sources. In other cases, especially those in which the proceeding has been set down for an adjudicatory hearing, the Commission staff has prepared a more traditional spread-sheet analysis. The spread sheet shows (1) the present value of the annual costs of constructing and operating the project over the term of the license, (2) the present value of the power sales revenues over the term of the license, (3) the net benefits or losses for each year, and (4) the cumulative net benefit or loss. Testimony of Robert Crowley in FERC Proceeding No. 4939 (available from coauthor Pope Barrow).

28. See Krutilla and Fisher, The Economics of Natural Environments (1975).
29. A very useful survey of the literature on economic analysis of river conservation and development is C. Dolcini and S. Anderson, River Valuation Bibliography: A Practitioner's Guide to River Valuation Literature (May 1986) (available from American Rivers).

Chapter 7

Other Tools for Protecting Rivers

The federal Wild and Scenic Rivers Act • Other types of federal river protection • The Clean Water Act • The federal land management agencies • States' regulation and influence • The Northwest Power Planning Council

WHILE FERC IS AT the center of the hydroelectric regulatory process, other agencies—at the federal and state levels—have important authority as well. In fact, some hydropower battles have been won outside of the FERC arena. (See No Way Big A on page 46.)

This chapter provides an overview of federal and state agency programs that influence hydroelectric development. It is intended to encourage you to pursue these alternatives; you can maximize your odds of succeeding by participating in these other forums. Since state programs vary and space allows only a thumbnail sketch of the federal programs, however, you will need to contact appropriate state and federal officials for more information.

FEDERAL WILD AND SCENIC RIVERS SYSTEM

Congressional designation of a *wild and scenic river* is the most effective and permanent way to protect a free-flowing river from development. The Wild and Scenic Rivers Act[1] prohibits FERC from approving a hydroelectric project either on or adversely affecting any designated river.[2] About two-tenths of a percent of America's river miles are now under wild and scenic protection.

Congress included 8 rivers in the wild and scenic rivers system when it adopted the act in 1968. Subsequent amendments to the act have added another 111 rivers, bringing the system's total size to 9,260 river miles. The system includes some of the most magnificent rivers in the country, including the Delaware, the Salmon, and the Tuolumne, to name but a few. There are active efforts underway to expand the system.

There are several ways that rivers can be added to the system. First, Congress can enact legislation officially designating a river as worthy of study for possible inclusion. Congressionally designated study rivers are off-limits to hydroelectric development while they are under evaluation. The study process is generally administered by federal officials, with assistance from local citizens and officials.

The National Park Service or the U.S. Forest Service (if the river is on Forest Service lands) prepares a report on the study river's eligibility and suitability for wild and scenic classification. The president then submits the report to Congress with a recommendation. If Congress accepts a recommendation in favor of designation and adopts the necessary legislation, the river is added to the official list of designated rivers.

Second, Congress can enact legislation adding a river to the system without first granting the river official study status. For example, as part of their ongoing resource planning process, the Forest Service and the Bureau of Land Management (BLM) have identified literally hundreds of rivers on their lands that are eligible for inclusion in the system. On their own initiative, these agencies are evaluating the suitability of wild and scenic status for these rivers, and they are expected to recommend such designation for many of them.

Finally, rivers can be included in the national wild and scenic rivers system through state initiative. Section 2(a)(ii) of the act recognizes that any state, with or without federal assistance, can identify potentially qualifying rivers within its boundaries and, under state law, develop conservation plans for them. The governor of the state may then submit a request to the secretary of the interior that the river be added to the system. The secretary must grant the request if it is determined that the river (1) has been designated as wild, scenic, or recreational by the state legislature, (2) will be permanently administered as a wild, scenic, or recreational river by the state or one of its political subdivisions, and (3) meets the basic criteria for inclusion in the federal system.

A state-designated wild and scenic river is not legally protected from inappropriate development until it has been added to the federal system under the section 2(a)(ii) process. However, once a river has been added to the federal system through such initiative, it has the same protection against hydroelectric development as a congressionally designated river.

Whatever designation process is followed, a river must possess certain characteristics to qualify for inclusion in the system. First, the river must be free-flowing—free of dams, channelizations, and other similar alterations. In addition, the river and its surrounding landscape must possess at least one outstandingly remarkable value in its scenic, recreational, geologic, fish and wildlife, historic, cultural, or other resources. Also, the act requires that each designated river be classified in one of three categories, according to the degree of development along the river. These categories are

1. wild—rivers that are free of impoundments and generally inaccessible except by trail, with watersheds or shorelines essentially primitive and waters unpolluted, representing "vestiges of primitive America"[3]
2. scenic—rivers that are free of impoundments, with shorelines or watersheds still largely primitive and shorelines largely undeveloped, but accessible in places by road[4]

3. recreational—rivers that are readily accessible by road or railroad, that may have some development along their shorelines and that may have undergone some impoundment or diversion in the past[5]

A management plan prepared for each river should seek to restrict development in or affecting the river corridor, to maintain the river's wild, scenic, or recreational character.

WILD AND SCENIC IN NEW ENGLAND

Wildcat Brook, one of the headwaters of the Saco River, is the scenic centerpiece of the town of Jackson, New Hampshire. In 1983 a Massachusetts power company submitted a preliminary permit application to FERC to study construction of a micro hydro project on famous Jackson Falls, in the center of the village. The town stopped the project—and obtained permanent protection for its river—by persuading Congress to include the brook in the national wild and scenic river system.

Aroused by the proposed hydro project, local officials and concerned citizens contacted American Rivers and others for their advice. This led to discussions with the National Park Service about the possibility of using the Wild and Scenic Rivers Act—which prohibits all new hydro development on designated rivers—to preserve this local treasure.

In June 1984, at the urging of the entire New Hampshire delegation, Congress passed a bill designating Wildcat Brook as an official wild and scenic study river. This step placed a six-year moratorium on all types of federally initiated or licensed water projects, including the proposed Jackson Falls project, that would adversely affect the value of the river and the adjacent landscapes. The National Park Service staff was then officially asked to initiate a comprehensive study.

The action on Wildcat Brook is a landmark in effective federal-local cooperation on a wild and scenic designation. Indeed, in enacting the federal study bill, Congress explicitly stated that the study should pursue wild and scenic protection relying on local government regulation and private landowner initiatives. Following enactment of the study bill, the National Park Service and the town of Jackson entered into a cooperative agreement to achieve this goal.

Following numerous public hearings, the town and the National Park Service developed a comprehensive river conservation plan. The primary implementing tools were zoning amendments eliminating future building in the floodplain, requiring building setbacks along the river, and establishing a new minimum lot size based on soil characteristics. In addition, the town, together with the Society for New Hampshire

Forests and the Wildcat River Trust, has acquired easements or fee interests in lands having high conservation interest; so far over 600 acres of land along the brook have been protected.

In traditional New England style, the citizens of the town gathered in a town meeting in March 1987 to adopt the zoning amendments. The vote in favor of the town's pursuit of a permanent wild and scenic designation was unanimous.

Congress passed the Wildcat Brook bill in the fall of 1988, and President Reagan signed the measure into law on 21 October 1988. Already, activists on other northeastern rivers, including the Pemigewasset in New Hampshire and the Genesee River in New York, are looking to Wildcat Brook as an example of an effective—and permanent—response to ill-advised hydro development.

FEDERAL LEGISLATIVE PROTECTION

In recent years Congress has relied increasingly on special approaches other than wild and scenic designation for dealing with rivers that present unique management problems.

Sharon McNeil

Wildcat Brook in New Hampshire. A locally initiated federal wild and scenic designation blocked a hydropower proposal in the town of Jackson.

One alternative is represented by the Columbia River Gorge National Scenic Area.[6] This national scenic area designation, adopted in 1986, provides for management and protection of an 85-mile stretch of the Columbia River and several hundred thousand acres of land in the river's gorge. The U.S. Forest Service, which owns much of the land in the gorge, administers the program together with a commission of representatives from the states of Washington and Oregon and from local counties. The legislation establishing the national scenic area also designates several tributaries of the Columbia River as wild and scenic rivers. Thus, the legislation incorporates and builds upon the Wild and Scenic Rivers Act but contains unique elements suited to the particular conditions and challenges in the Columbia River gorge.

Another federal approach is legislative veto of hydroelectric development that stops short of wild and scenic river designation. The Electric Consumers Protection Act (ECPA) contains a provision banning hydroelectric projects, with a few specified exceptions, on 61 miles of the Henry's Fork of the Salmon River in Idaho.[7] Unlike a wild and scenic designation, which is perpetual in nature, this law expressly contemplates that Congress may later lift the ban.

THE CLEAN WATER ACT

The Clean Water Act provides both the states and the U.S. Army Corps of Engineers a role in regulating hydroelectric development.

State Water Quality Certifications

Section 401 of the Clean Water Act authorizes the state water quality agency to veto proposed development inconsistent with attainment and maintenance of adequate water quality.[8] Any applicant for a federal license or permit that may result in a discharge into navigable waters must obtain state certification that the proposed activity will not result in a violation of federal or state water quality standards. This, of course, applies to those seeking licenses for hydroelectric projects. The state regulatory proceeding for section 401 certifications is an alternative forum for conservationists to oppose inappropriate hydroelectric development, or to obtain terms and conditions that will protect environmental resources. The 401 certification process applies both to original licenses and to those issued in relicensing proceedings.

The water quality standards applied in a section 401 proceeding are a blend of state and federal law. The Clean Water Act authorizes the states to adopt their own water quality standards and to enforce those standards through section 401, provided that the state standards are consistent with minimum federal requirements, established by the U.S. Environmental Protection Agency (EPA).

Project impacts of concern in a section 401 proceeding include changes in dissolved oxygen, temperature, turbidity, and other traditional measures of water quality. Other water uses, including fisheries and white-water boating, also have been treated as falling within the ambit of section 401. If the state pollution control agency is sympathetic to your cause, encourage the agency to exploit its authority under section 401 to the maximum extent possible.

Section 404 Discharge Permits

The U.S. Army Corps of Engineers also has independent regulatory control over hydroelectric development. The Clean Water Act requires a party to obtain a permit "for the discharge of dredged or fill material into the navigable waters at specified disposal sites."[9] Since the construction of any dam is regarded as a "discharge of fill material," the Corps has jurisdiction over virtually all new FERC projects.

For many years, there had been a question about whether the Corps' section 404 authority applied to hydroelectric projects. Some developers, as well as FERC, took the position that the Federal Power Act precluded the Corps from exercising jurisdiction over FERC projects. However, in *Monongahela Power Co. v. Marsh,*[10] a federal appeals court decided that FERC-licensed projects are not exempt from the requirements of the Clean Water Act, including section 404. It is now clear that hydroelectric developers must obtain development approval from both FERC and the Army Corps of Engineers.

In the Army Corps' review of an application, "the guiding principle should be that degradation or destruction of special sites may represent an irreversible loss of valuable aquatic resources."[11] Also, the guidelines state that, in general, "no discharge of dredged or fill material shall be permitted if there is a practicable alternative to the proposed discharge which would have less adverse impact on the aquatic ecosystem."[12] Under the Federal Power Act, the Commission has an obligation to consider the value of aquatic sites under the public interest test, discussed in chapter 3; under the Clean Water Act, the Corps has an express responsibility to protect these resources.

An Army Corps division or district engineer decides whether to grant a section 404 permit. Like the Commission, the Corps must comply with the National Environmental Policy Act (NEPA) in making its decision. Typically, where FERC and the Corps both have jurisdiction over a hydroelectric proposal, the agencies prepare a joint environmental analysis, with FERC acting as the so-called lead agency.

FEDERAL LAND MANAGEMENT AGENCIES

The federal land management agencies also have authority to regulate hydroelectric development on federal lands.

Under section 4(e) of the Federal Power Act, FERC decides whether issuing a license for a hydroelectric project on "reserved" federal lands will "interfere or be inconsistent with the purpose for which such reservation was created or acquired."[13] [A reservation is defined as a national forest, tribal lands within Indian reservations, military reservations, other federal lands withheld from appropriation under the public land laws (e.g., some BLM lands), and other federal lands acquired and held for any public purpose, but excluding national parks and national monuments.][14] Land management agencies such as the Forest Service and the Bureau of Land Management have authority only to prescribe conditions for project construction and operation that they "deem necessary for the adequate protection and utilization of such reservation."[15] In other words, the land management agencies can only condition—not veto—hydroelectric projects. Furthermore, their authority under section 4(e) of the Federal Power Act applies only to projects in federal reservations, so some federal agencies, such as the Bureau of Land Management, cannot condition projects on non-reservation lands.

The land management agency developing the section 4(e) conditions is responsible, under NEPA, for analysis specific to the project's impact on the federal reservation. Thus, even if FERC prepares an environmental impact statement on a project, the land management agency has its own responsibilities under NEPA.

Under the Federal Land Policy and Management Act (FLPMA), the Forest Service and the BLM can grant "rights of way" for "systems for generation, transmission, and distribution of electric energy."[16] There is an ongoing debate over whether this provision gives the Forest Service and BLM independent authority, like that of the Corps under section 404 of the Clean Water Act, to reject hydroelectric proposals. On its face, FLPMA appears to grant the land management agencies such veto power. But FERC has taken the position that the Federal Power Act was designed to give the Commission exclusive regulatory authority and that FLPMA does not alter this statutory design. In several orders, the Commission has expressly ruled that FERC licensees are not required to obtain rights of way from the land management agencies.[17] Despite FERC's position, many hydroelectric developers choose to comply with the agencies' requirements, rather than test the legal limits of their authority.

The local Forest Service and BLM offices decide how to test hydroelectric proposals for properties under their management. These decisions are then subject to appeal to these agencies' national offices. If you are concerned about a proposed project on federal lands, contact the supervisor of the national forest or the BLM area to discuss the environmental review process.

Since the enactment of ECPA, the Federal Power Act requires the Commission to consider the "extent to which a proposed project is consistent with a comprehensive plan (where one exists) for improving, developing, or conserving a waterway or waterways affected by the project" when a comprehensive plan has been prepared by an authorized agency.[18] Comprehensive plans include Forest Service and BLM resource management plans. Under this provision, FERC is required, absent compelling circumstances, to reject any hydropower proposal that conflicts with a recognized plan.

STATE RIVER PROTECTION

States concerned about the future of their free-flowing rivers are adopting programs and legislation to protect them. At present, 29 states have river protection programs. Some, with technical and financial assistance from the National Park Service, have prepared inventories or assessments of their river resources. Other states have enacted legislation creating regulatory programs to control unwise river development.

States have limited power to impose their conservation objectives on hydroelectric developers, because of the doctrine of federal preemption. Basically, when the federal government regulates a particular activity, it is up to Congress to define state power to regulate the same activity. As discussed earlier under section 401 of the Clean Water Act, states can enforce their water quality requirements against hydroelectric projects. Other opportunities for states to influence hydroelectric development are discussed here.

ECPA Provisions

ECPA requires FERC to consult with the states—and sometimes follow their direction—in determining whether or on what conditions to approve hydroelectric proposals.

First, FERC must consider comprehensive plans prepared by the state in which the facility is or will be located. A variety of state river protection programs, including state wild and scenic systems, qualify as comprehensive plans. As it does with federal management plans, FERC will generally follow the policies contained in state comprehensive plans.

Second, even in the absence of a comprehensive plan, the Commission must consider "the recommendations of . . . state agencies exercising administration over flood control, navigation, irrigation, recreation, cultural and other relevant resources of the State in which the project is located."[19] In addition, the Commission must consult with the state fish and wildlife agencies in accordance with the section 10(j) procedure.[20]

Finally, ECPA permits the states to determine whether projects receive benefits under the Public Utility Regulatory Policies Act (PURPA). States can, in effect, veto PURPA benefits for hydroelectric projects involving new dams and diversions unless the projects are consistent with state river protection policies. A developer constructing a new dam project cannot receive PURPA benefits if the project is located on (1) a "segment of a natural watercourse which is included in (or designated for potential inclusion in) a State . . . wild and scenic river system" or (2) a "segment of a natural watercourse which the State has determined, in accordance with applicable state law, to possess unique, natural, recreational, cultural, or scenic attributes which would be adversely affected by hydroelectric development."[21] Also, the state fish and wildlife agency can require that the developer of a PURPA project construct the project in accordance with terms and conditions to protect fish and wildlife.[22]

State Water Rights

Especially in the western United States, the states have administrative and judicial procedures for resolving conflicts due to competing uses of scarce water. By storing water or diverting it out of a stream, a hydroelectric project can have major effects on other existing or proposed water uses. Given federal preemption, it is uncertain whether the states can use their water laws to regulate hydroelectric development.

The Federal Power Act states that

> nothing contained in this chapter shall be construed as affecting or intending to affect or in any way to interfere with the laws of the respective States relating to the control, appropriation, use, or distribution of water used in irrigation or for municipal or other uses, or any vested right acquired therein.[23]

FERC asserts, however, that state control of the water needed for hydroelectric development would conflict with the Commission's planning and regulatory responsibilities.[24] According to FERC, the Federal Power Act ensures that, if a project interferes with an existing state water right, the owner of that right will be entitled to just compensation.

States take the position that the Federal Power Act preserved the states' authority to apply their water laws to hydroelectric projects in the same fashion as to any other water use.[25] This includes the power to reject a hydroelectric project when no unappropriated water is available under state law. It also includes the power to impose conditions to preserve state water resources by setting minimum stream flow requirements below a dam.

This is a debate over whether Congress intended to *preempt* state authority over hydroelectric development except when such regulation is expressly authorized by federal law, such as the Clean Water Act. In 1946, in *First Iowa Hydro-Electric Cooperative v. FPC,*[26] the U.S. Supreme Court ruled that the state of Iowa had no authority to require a FERC license holder to obtain a state permit to construct a hydroelectric dam. While First Iowa continues to be followed by the lower federal courts,[27] as well as by FERC, more recent Supreme Court decisions have undermined it.[28]

THE NORTHWEST POWER PLANNING COUNCIL

In 1980, Congress took an important step toward an intelligent balance between hydroelectric development and river preservation by enacting the Pacific Northwest Electric Power Planning and Conservation Act.[29] The act authorized the states of Idaho, Montana, Oregon, and Washington to join an interstate compact to create a regional policy-making body called the Northwest Electric Power and Conservation Planning Council, more commonly known as the Northwest Power Planning Council. The council focuses on two important Northwest resources—electric power and the Columbia River basin's fish and wildlife.

The council recently adopted a protected areas program. It identifies the remaining critical anadromous fish habitat in the region, as well as certain wildlife and high-value resident fish habitats, and sets them off-limits to hydropower. The Bonneville Power Administration, which is responsible for power distribution in the Pacific Northwest, will deny *intertie* access—access to the regionwide power grid—for power produced by dams on protected rivers in the Columbia River basin. In addition, FERC recognizes that the protected areas program is a type of comprehensive plan, which the Commission must consider in acting on hydroelectric proposals.

The comprehensive river planning effort of the council hopefully will provide a useful model for similar efforts in other states and regions of the country.

REFERENCES

1. See 28 U.S.C. §§ 1271–1287.
2. 16 U.S.C. § 1278.
3. 16 U.S.C. § 1273(b)(1).
4. 16 U.S.C. § 1273(b)(2).
5. 16 U.S.C. § 1274(b)(3).
6. Pub. L. No. 99–663.
7. ECPA, § 15A(a).
8. 33 U.S.C. § 1341.
9. 33 U.S.C. § 1344(a).
10. 809 F.2d 41 (D.C. Cir. 1987).
11. 40 C.F.R. § 230.1(d).
12. 40 C.F.R. § 230.10(a).
13. 16 U.S.C. § 797(e).
14. 16 U.S.C. § 796(2).
15. 16 U.S.C. § 797(e).
16. 43 U.S.C. § 1761(a)(4)
17. See, for example *In re Escondido Mutual Water Co.*, 9 FERC ¶ 61,241 (CCH) (26 Nov. 1979).
18. 16 U.S.C. § 803(a)(2)(A).
19. 16 U.S.C. § 803(a)(2)(B):
20. 16 U.S.C. § 803(j).
21. 16 U.S.C. § 824a–3(j)(2).
22. 16 U.S.C. § 842a–3(j)(3).
23. 16 U.S.C. § 821.
24. *In re Rock Creek Limited Partnership*, 38 FERC ¶ 61,240 (CCH) (11 Mar. 1987).
25. See Brief of State of California, *State of California v. FERC*, No. 87–7538 (filed 8 Apr. 1988 in the U.S. Court of Appeals for the Ninth Circuit).
26. 328 U.S. 152 (1946).
27. See *State of California v. FERC*, No. 87–7538 (6 June 1989, U.S. Court of Appeals for the Ninth Circuit).
28. See Whittaker, *The Federal Power Act and Hydroelectric Development: Rediscovering State Regulatory Powers and Responsibilities*, 10 Harv. Env. L. Rev. 135 (1986).
29. Pub. L. No. 96–501, 16 U.S.C. §§ 839–839h.

Chapter 8

Strategies for Effectively Dealing with Hydroelectric Projects

Strategic planning • Finding allies • Legal and technical experts • Raising money • The media

YOU MUST INTERVENE BEFORE FERC to deal effectively with a hydroelectric proposal. But you should expand your efforts beyond the filing of papers in a regulatory proceeding. This chapter provides some suggestions about how to develop that broader effort.

STRATEGIC PLANNING

Many citizens and groups with widely varying levels of interest, ability, and commitment are concerned about a typical hydroelectric proposal. The net will be less than the sum of the parts if each group pursues its own objectives without the benefit of a vision about where the cumulative effort is leading. Therefore, you need to prepare a strategic plan.

A key element of the plan must be a definition of the goals of the effort—where are we trying to get to from here? Sometimes the goal will be obvious, such as protecting a pristine stretch of free-flowing river from a dam. Sometimes the goal will be more limited; for example, citizens may participate in a relicensing proceeding to restore trout habitat or to obtain releases for white-water boating for a specific number of days each year.

One goal should be to avoid having to fight the same battle again a few years later. If a river is worth defending against a hydroelectric proposal, then it should be protected permanently through inclusion in the federal or a state wild and scenic river system or through private conservation easements. A simultaneous campaign to prevent or limit development and to obtain permanent protection of a river makes effective use of scarce conservation resources. It also has a positive tone, which is helpful in seeking public and political support.

Plot the steps necessary to achieve your ultimate goal. In addition to preparing and filing an intervention with FERC, you should consider participating in state proceedings related to the project or consulting with state officials about the project. A successful campaign may involve press contacts, fund raising, obtaining the support of politicians, and so on. A strategic plan is a way to prioritize what needs to be done and to assign different tasks to appropriate members of your group.

As the campaign proceeds, you will need to adjust the plan. Be flexible, but always know where you are headed.

FINDING ALLIES

Allies—especially powerful ones—increase your chances for success. Allies are useful in the FERC proceeding—for sharing the work and for demonstrating that your position has broad support. Allies can be useful in taking the many other steps necessary to preserve a river, as well.

Find others to join with you as parties in the FERC proceeding. Other individuals and nonprofit organizations either can file their own motions to intervene or can join in your motion. Interventions are commonly filed on behalf of as many as a dozen individuals and groups. (Such a motion must explain why each member of the coalition is entitled to participate under the Commission's rules.) Encourage state and federal agencies concerned about a project to file their own interventions.

Try to enlist allies who will add weight to the motion. The Commission is likely to give greater consideration to the views of an established organization than to those of an individual. A coalition that includes different groups (for example, boaters, anglers, outfitters, chambers of commerce) will carry more weight with the Commission than a coalition of groups dedicated to a single purpose. If you cannot convince allies to intervene, ask them to at least write letters to FERC voicing support for your position.

Allies are useful outside the context of FERC, also. In a number of instances, a hydroelectric battle ultimately has been won in the state legislature or in Congress. Hydroelectric proposals always have to undergo at least some type of state review; allies can be valuable in the state proceedings just as they are in a FERC proceeding. A federal wild and scenic designation, which prohibits hydroelectric development, is an option for blocking a proposed project on a free-flowing river; many allies obviously are needed in the federal arena to generate political support for a legislative proposal.

Potential allies include local groups and individuals who enjoy the river and are already committed to seeing it preserved. Contact such groups. American Rivers and other national groups also can help you identify individuals and groups in your state or region that are knowledgeable about hydropower issues.

Try to enlist national conservation groups as allies in a proceeding. Their participation signals to FERC the importance of the proceeding. Also, some national groups, including American Rivers, the American Whitewater Affiliation, Friends of the Earth, the Izaak Walton League, the National Audubon Society, the National Wildlife Federation, the

Sierra Club, and Trout Unlimited, have extensive expertise in analyzing and opposing hydropower proposals. (See Appendix B for the names and addresses of national conservation groups interested in hydropower issues.)

Use your imagination in seeking out allies. Proponents of economic development are potentially valuable allies, especially in communities where commercial boating depends upon preservation of the river. In a heated battle in the early 1980s over a hydroelectric proposal on the Gauley River in West Virginia, the local chamber of commerce took the lead in opposing the project in order to protect white-water boating. In another instance, the United Mine Workers opposed a project because of its concern about the potential impact of hydropower development on jobs in the coal industry. Ranchers or others whose water rights are threatened by a hydropower proposal can be active allies, especially in the western United States.

Political figures can be helpful. A letter to the Commission from a U.S. Senator or Congressperson has proven effective in the past in convincing the Commission to prepare an environmental impact statement on a project. The Commission is sensitive to the concerns of state and local politicians, as well. Consider asking the city council or county board to pass a resolution opposing the project or requesting a thorough environmental review.

Including large numbers of individuals and groups in your coalition can present a challenge. The coalition can become too diffuse and unmanageable. If many groups with different interests join in a single motion to intervene, confusion and conflict can result. A coalition is obviously counterproductive if it is so large or includes so many conflicting interests that it has difficulty meeting or reaching a consensus.

LEGAL AND TECHNICAL EXPERT ASSISTANCE

The Importance of Experts

A citizen or group intervening in a hydroelectric proceeding can usually benefit from expert legal help. The Commission's emphasis on procedure can be a major disadvantage for an inexperienced or naive intervenor, but it can be an advantage to a sophisticated intervenor with adept legal counsel. Procedural arguments are part of the process at the Commission, and you must use them to your advantage whenever possible.

In addition to being procedurally complex, Commission proceedings also can be substantively complex, especially if they raise difficult questions about project feasibility or need for power. Furthermore, the Commission focuses hard on what evidence each party has introduced into the record in each proceeding. A professional economist or engineer can be indispensable in developing and presenting the evidence you need to make your case persuasive.

Finding a Lawyer

There are many ways to find a good lawyer to assist you. A lawyer in your community who already is committed to the cause may be willing to represent you. While prior experience in dealing with FERC is helpful, it is by no means essential: any competent attorney can grasp the Commission's procedures and can read the leading court decisions.

A major law firm may provide assistance in a hydroelectric proceeding, especially if you can locate individuals within the firm who are interested in river protection. Some firms in large cities have institutionalized the lawyers' professional obligation to provide *pro bono*—free—assistance to individuals and groups that otherwise could not afford representation. These firms expressly allow, and sometimes encourage, their lawyers to take on *pro bono* matters. If you do not have a specific contact at a firm, you can make a blind call and ask to speak with the lawyer in charge of *pro bono* assignments. Understand, however, that the firm will not be able to represent you if that would conflict with the interests of its existing clients, including industrial or financial companies.

You also may be able to obtain legal assistance through national conservation organizations, some of which employ lawyers who have experience in hydropower matters. Especially if your proceeding involves a river of national importance, or if it raises a novel legal issue, such organizations may be willing to prosecute the case on behalf of a broad coalition of environmentalists. National conservation organizations may be able to identify individual lawyers or other conservation groups that can assist you. The American Whitewater Affiliation and American Rivers maintain a "river watch" file of potential volunteer lawyers in different regions of the country.

Technical Help

It also is useful to obtain technical assistance to help you analyze a project's economic or engineering feasibility and its environmental impacts. The expertise that you need depends on the specific issues raised by your proceeding.

Hiring technical experts at market rates can be very expensive. However, some experts may be willing to charge reduced fees if they believe you have a worthy cause and lack the resources to pay market rates. Also, the memberships of local conservation organizations commonly include technical experts who may be willing to donate their services. The staffs of local, state, or national conservation groups, or professors and researchers at local colleges and universities, also may be willing to volunteer their help.

There are many ways to find an expert. Professional consulting firms, colleges and universities, and conservation organizations may be able to offer assistance or put you on the trail of an appropriate expert. The "river watch" file maintained jointly by the American Whitewater Affiliation and American Rivers lists technical experts interested in hydropower issues.

For better or worse, the professional credentials of your experts carry weight. Select an expert with an educational background and professional credentials that demonstrate competence.

RAISING MONEY

A hydroelectric campaign costs money. Even preparing and filing a simple FERC intervention can cost several hundred dollars for typing, copying, telephone calls, and mailing. The national campaigns to stop hydroelectric projects on the Tuolumne River in California and on the West Branch of the Penobscot in Maine cost several hundred thousand dollars apiece.

There are a few special twists to fund-raising for river conservation. A hydroelectric project that will dam a free-flowing river, destroying important recreational or wildlife resources, is a concrete threat to the environment that attracts public support. Your fund-raising efforts should emphasize the immediacy of the threat.

The river itself can be used as a fund-raising resource. Commercial outfitters may contribute raft or canoe trips for a benefit, provide publicity for the river campaign, sell raffle tickets, and so on. Their livelihood depends on maintaining the free-flowing character of the river.

A river festival is another option. A typical river festival, like the Gauley River Festival in West Virginia or the Ocoee River Festival in Tennessee, is a daylong affair with a variety of boating activities, entertainment, and enthusiastic talk. A successful festival can serve not only as a fund-raising tool, but also as a public relations event. The major drawback to a festival is the large up-front cost for the site, food, entertainment, tickets, and parking, along with the risk that the outlay may not be recouped if the festival is unsuccessful.

Raffles are another option. Manufacturers and suppliers of outdoor equipment can be generous in donating canoes, river trips, fishing gear, tents, and similar items for use in raffles to raise money for river campaigns.

In general, fund-raising through a direct mailing is an expensive and risky way to raise money. There are high up-front costs for printing, postage, and the purchase of lists, and the returns are unpredictable. Face-to-face or telephone requests for donations, while somewhat more difficult to administer, are safer and potentially more successful techniques. The Savage River Defense Fund, a group organized in 1987 to fight a hydro proposal in Maryland, raised $15,000 through phone solicitations in a few months.

FAVORABLE PRESS COVERAGE

Press coverage helps determine whether your conservation campaign will succeed. While the Commission is less responsive to publicity than are nonadjudicatory agencies, press stories that emphasize the values of the river can exert a subtle influence on the Commission's thinking.

Press coverage will influence the level of public and political support you achieve. Good press can be a stimulant for your supporters and can encourage others to join you. Bad press can be demoralizing and hinder your efforts to gain support.

There is a variety of useful techniques for gaining favorable press attention. The following is a partial list of techniques that have proven useful in the past.

- Get to know the reporter who will be covering the story. Your best assurance of good coverage is a reporter who likes you, has a genuine interest in the subject, and above all else, has come to rely on you as a trustworthy source of information.
- The press loves an event on which to hang a story. A visit to the river by a celebrity or a prominent political figure is a perfect opportunity to garner press coverage. Any fund-raising event can do double duty as a press event. Local press give a river conservation battle greater play when national conservation groups join the fray. American Rivers' list of the ten most endangered rivers or the American Whitewater Affiliation's Hydromania Award is a perfect peg for positive press coverage.
- A short, well-written press release is an effective way to inform the press about your activities and to generate helpful coverage, especially if you send the press release after an initial phone call.

- The press—especially overworked local reporters—love to be spoon-fed. Your press releases should be one page long, if possible, and written in newspaper style. Also, newspapers and magazines frequently will agree to print articles you write explaining your position, especially if you do not ask to be paid for the piece.

THE HYDROMANIA AWARD

"The American Whitewater Affiliation hereby presents the annual Hydromania Award to Western Power, Inc. in recognition of Western Power's misguided efforts to construct the GEM Irrigation District Power Project (#10396), an environmental insult of national magnitude which would result in extensive and permanent damage to the magnificent natural and recreational resources of the North Fork of the Payette River in Idaho."

This award was presented in January 1989 to publicize a disastrous project under consideration in Idaho. Bad publicity may not kill an ill-advised hydro project, but it can make government bureaucrats and politicians wary about giving it the green light.

The proposed project would be huge (350 megawatts) and ruinous for a famous white-water stream. Western Power proposed to build a dam and two diversions, which would eliminate about 24 miles of white water, including a 16-mile class 5 section known throughout the world as the Mt. McKinley of white water. The dam would flood 6,700 acres of ranch land and homesites. Since Idaho has a huge power surplus, project backers would build transmission lines to Nevada and southern California to market the power.

In 1988 the American Whitewater Affiliation was searching for a way to heighten public awareness about the Idaho project. As usual, most people knew nothing about it. With John Q. Public asleep at the switch, American Whitewater activists feared the developer would make too much headway before opposition developed. After sinking huge sums in planning, the developer could become stubbornly entrenched. Opposition might be portrayed as a last-minute ambush.

Project opponents needed the press to wake up the public. They wanted the developer, the state and federal agencies, and the press to know—in advance—that the project was environmental trouble. But they had a problem. There was no news event—nothing to generate publicity. Meanwhile, the company had high-powered consultants quietly working behind closed doors, drawing up elaborate plans. Everything would be handled by lawyers plowing through mounds of incomprehensible paperwork in obscure offices in Washington, D.C. Bulldozers and concrete mixers would not show up on the river for several years, but if conservation groups waited that long, it would be too late.

Reminded of "lemon" awards like the Golden Fleece or the Dirty Dozen, American Whitewater activists hit on a Hydromania Award. It would call attention to the project and focus opposition. It would shine a spotlight on the proposal. Another plus—it could be a lot of fun.

After careful preparation and after numerous calls to newspaper reporters, television stations, and other media outlets, American Whitewater activists traveled to Washington for a press briefing. Media people and political leaders were invited.

At the event itself, while reporters took notes and TV cameras rolled, American Whitewater leaders read the award letter and opened the floor to a full-scale question and answer session. At the same time, copies of the announcement were delivered to key decision makers in the Idaho state legislature, the U.S. Congress, and the Federal Energy Regulatory Commission.

Wendy Wilson

The North Fork Payette. The hydropower developer who targeted this outstanding river was honored with American Whitewater Affiliation's Hydromania Award.

Appendix A

FERC Addresses and Telephone Numbers

D.C. Headquarters
Federal Energy Regulatory Commission
825 North Capitol Street, NE
Washington, D.C. 20426
(202) 357–5200

Key Offices

Public Information Room—Room 1000
(202) 357–8118

Office of the Secretary
(202) 357–8400

Office of Hydropower Licensing
(202) 376–9171

Regional Offices

Atlanta Regional Office
Federal Energy Regulatory Commission
730 Peachtree Street, NE
Room 800
Atlanta, Georgia 30308
(404) 347–4134

Chicago Regional Office
Federal Energy Regulatory Commission
230 South Dearborn Street
Room 3130
Chicago, Illinois 60604
(312) 353–6173

New York Regional Office
Federal Energy Regulatory Commission
201 Varick Street
Room 664
New York, New York 10014
(212) 337–2609

Portland Regional Office
Federal Energy Regulatory Commission
1120 South West Fifth Avenue
Suite 1340
Portland, Oregon 97201
(503) 294–5840

San Francisco Regional Office
Federal Energy Regulatory Commission
Third Floor
San Francisco, California 94103
(415) 974–7150

Appendix B

Conservation Groups Concerned About Hydropower Issues

American Rivers
801 Pennsylvania Avenue, SE
Suite 303
Washington, D.C. 20003
(202) 547–6900

American Whitewater Affiliation
c/o Risa Shimoda
426 Page Avenue
Number 11
Atlanta, Georgia 30307

Environmental Defense Fund
257 Park Avenue South
New York, New York 10010
(212) 505–2100

Environmental Policy Institute / Friends of the Earth
218 D Street, SE
Washington, D.C. 20003
(202) 544–2600

Friends of the River
909 12th Street
Suite 207
Sacramento, California 95814
(916) 442–3155

The Izaak Walton League of America
1701 North Fort Myer Drive
Suite 1100
Arlington, Virginia 22209
(703) 528–1818

National Audubon Society
801 Pennsylvania Avenue, SE
Washington, D.C. 20003
(202) 547–9009

National Wildlife Federation
1400 16th Street, NW
Washington, D.C. 20036–2266
(202) 797–6800

Natural Resources Defense Council
122 East 42nd Street
New York, New York 10168
(212) 949–0049

Save Our Streams
P.O. Box 56
North Fork, California 93643
(209) 877-2978

Trout Unlimited
501 Church Street, NE
Vienna, Virginia 22180
(703) 281–1100

Appendix C

Model Motion to Intervene

UNITED STATES OF AMERICA
FEDERAL ENERGY REGULATORY COMMISSION

In re Intermountain Power Corporation
Oxbow Bend Hydroelectric Project

Project No. 6329–001

MOTION OF AMERICAN RIVERS, AMERICAN CANOE ASSOCIATION, AND AMERICAN WHITEWATER AFFILIATION TO INTERVENE IN OPPOSITION; REQUEST FOR PREPARATION OF ENVIRONMENTAL IMPACT STATEMENT; AND REQUEST FOR ADJUDICATORY HEARING

American Rivers, American Canoe Association, and American Whitewater Affiliation (hereafter "intervenors") hereby move for leave to intervene in opposition in the above-captioned proceeding. Intervenors further request that the Commission direct the staff to prepare an environmental impact statement and conduct an evidentiary hearing to resolve disputed issues of fact relating to the proposed project. In support of this motion and these requests, intervenors state as follows:

INTRODUCTION

The proposed Oxbow Bend hydroelectric project would be located on the South Fork Payette River approximately four miles west of Lowman in Boise

County, Idaho. The major project facilities, all of which would be located on public Forest Service lands, would consist of a 30-foot-long, 10-foot-high diversion dam; an 850-foot-long, 12-foot-diameter buried penstock; and a powerhouse containing two generating units with a total capacity of approximately three megawatts. The project would result in the diversion of a maximum of 1,500 cfs from the river. A stretch of river approximately one mile in length would be significantly dewatered if the project were built.

By any standard, the South Fork Payette River is an outstanding river. The Nationwide Rivers Inventory, prepared by the National Park Service, includes the South Fork Payette among the mere 2 percent of the nation's rivers identified in the inventory as deserving of consideration for inclusion in the federal wild and scenic rivers system. The Idaho Department of Fish and Game has stated that the South Fork Payette "is classified as Class I, or having the highest valued fishery resource, in a joint U.S. Fish and Wildlife Service, U.S. Environmental Protection Agency and Idaho Department of Fish and Game stream evaluation" [Motion to Intervene, at 2 (filed 1 May 1987)] and that the river is "contained in the mid-South Fork Payette River drainage that, in total, comprises one of the most popular and heavily fished stream areas in the state of Idaho" (Letter from Mr. Conley to Mr. Auth, dated 15 April 1982). The South Fork Payette also is recognized as one of the most significant and popular white-water boating and rafting rivers in the region.

Because the costs of the project, based on the information currently available, clearly outweigh the potential benefits, intervenors are opposed to the project. Available evidence in the record indicates that the downstream releases will very likely be inadequate to maintain the high-quality trout and whitefish fishery. In addition, it appears to be uncontroverted that the project will essentially destroy the approximately one-mile stretch below the intake structure as a white-water boating resource. On the other hand, there is no evidence of any need for the power the project would generate; the most recent figures prepared by the Western Systems Coordinating Council indicate that, in the Northwest Power Pool, existing and already planned additions to capacity will provide a substantial excess power supply in the Northwest Power Pool through at least the year 1996. In addition, the application provides no reliable information about the economic viability of the proposed project, especially in light of the most recent proposed modifications to the project's operations.

STATEMENTS OF INTEREST
AMERICAN RIVERS

American Rivers is a not-for-profit conservation organization with offices located at 801 Pennsylvania Avenue, SE, in Washington, D.C. The mission of American Rivers is to preserve the nation's outstanding rivers and their landscapes. With approximately 9,000 members across the country, American Rivers is the nation's principal river conservation organization. Approximately 100 of American Rivers' members reside in Idaho.

American Rivers is entitled to intervene in this proceeding because its members use and enjoy the South Fork Payette River for fishing, white-water boating, and other forms of recreation. Since the proposed project would significantly impair the value of the South Fork Payette for these activities, American Rivers' members have a direct and immediate interest in whether or not the project is constructed.

American Rivers' participation in this proceeding also is in the public interest. As a national organization, American Rivers' participation will ensure that the value of the South Fork Payette to fishermen and boaters who reside in all parts of the country is properly reflected in this proceeding. In addition, because American Rivers has recognized expertise in the preservation of wild and scenic rivers and in the law and policy concerning hydroelectric development, American Rivers will bring distinct expertise to the difficult and controversial issues raised by this proceeding.

AMERICAN CANOE ASSOCIATION

The American Canoe Association (ACA) is a nonprofit corporation organized under the laws of the state of New York on 7 November 1927. It maintains its principal mailing address at 8580 Cinderbed Road, Suite 1900, P.O. Box 1190, Newington, VA 22122–1190. The general purpose of the ACA is "to unite all persons interested in canoeing, and . . . provide educational, informational, and training services to increase the safety, enjoyment and skills of those canoeing on rivers, lakes, estuaries and other bodies of water."

The ACA has a membership of approximately 5,000 nationwide, representing many of the most dedicated and active individuals engaged in canoe sport in this country. ACA members use and enjoy the South Fork Payette River for recreation, and therefore they have a direct and immediate interest in this proceeding.

AMERICAN WHITEWATER AFFILIATION

The American Whitewater Affiliation (AWA) is a Missouri not-for-profit corporation. The AWA maintains its principal mailing address at 146 North Brockway, Palatine, IL 60067. The AWA was organized to protect and enhance the recreational enjoyment of white-water sports in America. Since its incorporation in 1961, the primary focus of the AWA has been the preservation, protection, and enjoyment of white-water rivers in America and throughout the world.

The current membership of the AWA includes approximately 1,300 individual or business members and 61 affiliate members. The affiliate members are canoe clubs throughout the nation whose members are interested in the white-water aspects of canoe sport. The sizes of clubs vary from small clubs with less than 100 members to large clubs such as the Canoe Cruisers Association with more than 1,300 individual members. Members of the AWA use and enjoy the

South Fork Payette River for white-water sports. Their interests will be adversely affected by the proposed project.

I. NO LICENSE CAN ISSUE UNTIL THE SUITABILITY OF THE SOUTH FORK PAYETTE FOR INCLUSION IN THE FEDERAL WILD AND SCENIC RIVERS SYSTEM HAS BEEN DETERMINED.

Section 5(d) of the Wild and Scenic Rivers Act (Pub. L. No. 95–542, 16 U.S.C. §1271 et seq.) requires all federal agencies to consider potential national wild, scenic, or recreational rivers in all planning for the use and development of water and related land resources. The responsibility imposed by section 5(d) plainly requires the Commission to assess potential wild and scenic rivers in conducting comprehensive planning under section 10 of the Federal Power Act. Similarly, the U.S. Forest Service has a responsibility to study potential additions to the federal wild and scenic rivers system during the preparation of land and resource management plans pursuant to the National Forest Management Act.

Recognizing that federal agencies "need to speed up the process for studying Wild and Scenic Rivers for designation," President Carter in his 1979 Message to Congress on Environmental Priorities and Programs directed federal land management agencies to assess whether rivers located on their lands and identified in the Nationwide Rivers Inventory prepared by the Heritage Conservation and Recreation Service (subsequently published by the National Park Service) are suitable for inclusion in the wild and scenic rivers system. The president ordered federal agencies to "take prompt action to protect [suitable] rivers—either by preparing recommendations for their designation or by taking immediate action to protect them." The president's Memorandum for the Heads of Departments and Agencies, dated 2 August 1979, reiterated these instructions and specifically ordered agencies, to the extent of their authority, to "promptly take such steps as are needed to protect and manage [each suitable] river and the surrounding area in a fashion comparable to rivers already included in the Wild and Scenic Rivers System."

The South Fork Payette River was included in the Nationwide Rivers Inventory, and the Forest Service has specifically identified the South Fork Payette as a potential wild, scenic, or recreational river. (See Letter to Secretary Plumb from Regional Forester J. S. Tixier, filed 23 May 1985.) In light of the mandates discussed above, the Commission and/or the Forest Service have the responsibility to assess the suitability of the river for inclusion in the federal wild and scenic rivers system before acting upon this development proposal. In view of the fact that the Forest Service is responsible for managing the land area affected by the proposed project, and is in the process of completing a forest plan for the area, it is most appropriate for the Forest Service to conduct the suitability analysis. In any case, licensing action on this project must await completion of the suitability determination.

The applicant is wrong in contending that the construction of this project would not be inconsistent with the future inclusion of the river in the federal wild and scenic rivers system. While the applicant contends that the relevant stretch of the South Fork Payette River would qualify, at best, for designation as a recreational river, so far as we are aware, no classification has yet been made. At least until the Forest Service has made a classification, the Commission and the Forest Service should continue to manage the river in order to protect its present values. (See chapter 8 of the Forest Service's *Land and Resource Management Planning Handbook,* section 8.12.)

Furthermore, contrary to the applicant's position, even if the river were classified as a potential recreational river, chapter 8 of the Forest Service's *Land and Resource Management Planning Handbook* prescribes specific management guidance for rivers, like the South Fork Payette, that have been identified for further study. These guidelines proscribe development of hydroelectric power on potential wild, scenic, *or* recreational rivers.

Accordingly, apart from the merits of this particular project, no action can be taken on this license application until the suitability of the South Fork Payette River for inclusion in the federal wild and scenic rivers system has been resolved.

II. THE PROPOSED PROJECT WILL HAVE UNREASONABLE ADVERSE EFFECTS ON IMPORTANT NATURAL RESOURCES

The available evidence demonstrates that the proposed project will have serious adverse effects on the environment.

For example, it appears to be undisputed that the proposed project will seriously degrade the value of the South Fork Payette for white-water boating. The proposed stipulation between the applicant and the Idaho Department of Parks and Recreation, filed with the Commission on 16 October 1986, only provides for a guaranteed bypass flow of 300 cfs from Monday through Thursday and 400 cfs from Friday through Sunday and on holidays, to be maintained from Memorial Day weekend through Labor Day weekend. The stipulation expressly acknowledges that this bypass flow is not designed to maintain a high-quality boating experience; the stipulation states that the applicant and the Commission simply intend "that floatboats be able to safely float through the project reach and, should that prove undesirable under certain situations, be able to launch immediately downstream of the project."

Ironically, the person presented by the applicant as an expert on white-water boating appears to confirm the project's serious adverse threat to white-water boating. (See Letter of Stephen J. Guinn, filed on 15 April 1985.) Mr. Guinn states that "an intermediate float boater" should have "no problem" at the affected stretch of the South Fork Payette "in water flows from 1600 to 400 cfs." He then goes on to state, however, "that a desirable flow for boating would be between 900 cfs to 1400 cfs."

Similarly, the proposed project appears to present unacceptable adverse effects on the fishery. The applicant is proposing to maintain a minimum bypass flow of 300 cfs. However, both the Idaho Department of Fish and Game and the U.S. Fish and Wildlife Service have taken the position that a higher minimum flow will be necessary to maintain the existing high-quality trout and whitefish fishery. (See Letter to Mr. Auth from Directory Conley and Letter to Mr. Auth from Field Supervisor Wolflin, filed with the Commission on 8 October 1986.)

III. THERE HAS BEEN NO SHOWING OF A NEED FOR POWER

So far as we can determine, the applicant has presented no evidence that there is any actual need in the relevant region for the electricity that the proposed project would generate.

Authoritative estimates of future power needs in the northwestern United States indicate that there is not likely to be any need for the power the project would generate until late in the next decade if not beyond. For example, information compiled by the Western Systems Coordinating Council demonstrates that existing and already planned additions to the region's generating capacity will result in a substantial excess generating capacity over peak power demands at least through 1996. (See Attachment A.) The council estimates that in 1996 the margin during the summer will be 3,055 megawatts and that the margin during the winter will be 3,418 megawatts. Thus, the construction of the proposed project would simply exacerbate an existing power surplus in the northwestern United States.

IV. THERE IS A LACK OF EVIDENCE OF ECONOMIC FEASIBILITY

The applicant has not provided any relevant information demonstrating that the project is economically viable. In light of the importance of the natural resources threatened by the project, a showing of economic feasibility is a prerequisite for issuance of the license.

Despite the absence of detailed information, certain aspects of the application raise a serious question about the project's viability. First, it appears that the applicant has made a number of modifications in the proposed construction and operation of the facility over the last several years; yet there is no evidence that the applicant has analyzed the costs of these modifications to determine their effect on the economic viability of the project.

Second, in its filing with the Commission on 8 October 1986 (at 8), the applicant states that the difference between the 300 cfs minimum flow recommended by the applicant and the 337 cfs minimum flow recommended by fisheries agencies "is of importance to the project's economics and viability, because the 37 cfs disparity is year around base flow which, otherwise, may not be available for generation." If, as this statement suggests, a difference in the bypass flow of 37 cfs could make the difference between the viability and nonviability of the project, this raises at least a substantial question as to whether the project is, at best, marginally viable.

V. THE COMMISSION SHOULD DIRECT THE STAFF TO PREPARE AN ENVIRONMENTAL IMPACT STATEMENT ON THE PROPOSED PROJECT

The National Environmental Policy Act (NEPA) requires that an environmental impact statement (EIS) be prepared for "major federal actions significantly affecting the quality of the human environment" (42 U.S.C. § 4332(2)C). The courts have consistently held that an EIS must be prepared when a proposed federal action may have significant effects on recreational and fishery resources. Furthermore, where "substantial questions" have been raised about whether a proposed hydropower development "may have a significant effect, an EIS *must* be prepared" (emphasis in original) *The Steamboaters v. FERC*, 759 F.2d 1382, 1392 (9th Cir. 1985).

The regulations of the Council on Environmental Quality (CEQ) set forth detailed criteria for when federal agencies must prepare an environmental impact statement. Two criteria are of particular relevance in this case: (1) "unique characteristics of the geographic area such as proximity to historic or cultural resources, park lands, prime farm lands, wetlands, wild and scenic rivers, or ecologically critical areas" and (2) "the degree to which the effects on the quality of the human environment are likely to be highly controversial" (40 C.F.R. § 1508.27).

There cannot be any dispute that the proposed project area has unique characteristics. The stretch of the South Fork Payette that would be affected by the project has been identified as among the finest fishing and boating rivers in the state of Idaho. Furthermore, the river has been identified in the Nationwide Rivers Inventory as worthy of inclusion in the 2 percent of the nation's rivers that deserve study for designation as part of the federal wild and scenic rivers system.

Furthermore, the record in this proceeding clearly demonstrates that the proposed project and its likely impacts have become highly controversial. At least one federal and one state agency have expressly asserted their opposition to the project as currently designed. The applicant's views that the project will not adversely affect the potential for including the river in the federal wild and scenic rivers system, and that the project will not adversely affect the fishery, are contradicted by expert government agencies. In addition, numerous individuals and groups have specifically notified the Commission of their concern about the proposed project.

In sum, NEPA, controlling precedent, and applicable CEQ regulations all require the preparation of an EIS on the Oxbow Bend project.

VI. AN EVIDENTIARY HEARING MUST BE HELD ON THE PROJECT BEFORE A LICENSE CAN BE ISSUED

As indicated by the discussion above, there are numerous genuine issues of material fact that are in dispute in this proceeding. Among others, these include:

- whether construction of the proposed project would be inconsistent with the future designation of the South Fork Payette River as a recreational river
- whether the project will have serious adverse effects on the quality of the environment, including the existing fishery and recreational opportunities
- whether there is need for the electric power the project would generate
- whether the project is economically viable

Intervenors are prepared to offer, along with other intervenors, evidence and testimony, including evidence from experts on the value of the South Fork Payette as a potential wild and scenic river, on power needs in the northwestern United States, and on the recreational importance of the South Fork Payette.

CONCLUSION

For the foregoing reasons, the Commission should grant the motion to intervene, direct the preparation of an environmental impact statement, and order the proceeding set for a hearing.

Respectfully submitted,

____

John D. Echeverria, Esquire
American Rivers, Inc.
801 Pennsylvania Avenue, SE
Washington, D.C. 20003
(202) 547–6900

____

Dr. Robert Kauffman
ACA National Conservation Chairman
Dept. of Recreation
HPER Bldg.
University of Maryland
College Park, Maryland 20742
(301) 454–3383

____

Pete Skinner
Director of AWA
Box 272
Snyder Road
West Sand Lake, New York 12196
(518) 587–1204

16 November 1987

CERTIFICATE OF SERVICE

I hereby certify that I have this 16th day of November 1987 served the foregoing document upon each person designated on the official service list compiled by the secretary in this proceeding.

____

John D. Echeverria

ATTACHMENT A

[Western Systems Coordinating Council, COORDINATED BULK POWER SUPPLY PROGRAM 1986–1996]

Item 3–A
Estimated Peak Resources, Demand and Margin
For the 1 to 10 Year Period

Council-WSCC
U.S. Systems

		1987		*1988*		*1989*		*1990*		*1991*	
	Resources in MW	*Summer*	*Winter*	*Summer*	*Winter*	*Summer*	*Winter*	*Summer*	*Winter*	*Summer*	*Winter*
01	Net Dependable Capability	123087	126124	126908	127831	128761	130181	130145	130497	130451	130774
02	All Scheduled Imports	868	800	843	775	723	755	723	570	638	670
03	All Scheduled Exports	200	200	200	200	200	0	0	0	0	0
04	Total Resources (01+02−03)	123755	126724	127551	128406	129284	130936	130868	131067	131089	131444
	Demand in MW										
05	Peak Hour Demand	85932	82040	87608	83904	90011	86015	91764	87826	93690	89523
06	Interruptible Demand	1513	1340	1645	1313	1688	1327	1772	1398	1770	1375
07	Demand Requirements (05−06)	84419	80700	85963	82591	88323	84688	89992	86428	91920	88148
	Margin in MW										
08	Margin (04−07)	39336	46024	41588	45815	40961	46248	40876	44639	39169	43296

		1992		1993		1994		1995		1996	
	Resources in MW	*Summer*	*Winter*	*Summer*	*Winter*	*Summer*	*Winter*	*Summer*	*Winter*	*Summer*	*Winter*
01	Net Dependable Capability	130489	131298	131177	131917	132458	133381	133688	134434	134372	134765
02	All Scheduled Imports	638	670	638	670	613	645	688	720	663	434
03	All Scheduled Exports	0	0	0	0	0	0	0	0	0	0
04	Total Resources (01+02−03)	131127	131968	131815	132587	133071	134026	134376	135154	135035	135199
	Demand in MW										
05	Peak Hour Demand	95645	91274	97532	93180	99604	94901	101715	96960	103604	97978
06	Interruptible Demand	1763	1367	1763	1379	1785	1405	1804	1412	1795	1353
07	Demand Requirements (05−06)	93882	89907	95769	91801	97819	93496	99911	95548	101809	96625
	Margin in MW										
08	Margin (04−07)	37245	42061	36046	40786	35252	40530	34465	39606	33226	38574

Item 3–A
Estimated Peak Resources, Demand and Margin
For the 1 to 10 Year Period

Council-WSCC
Party-Northwest Power Pool
Area

		1987		*1988*		*1989*		*1990*		*1991*	
	Resources in MW	*Summer*	*Winter*	*Summer*	*Winter*	*Summer*	*Winter*	*Summer*	*Winter*	*Summer*	*Winter*
01	Net Dependable Capability	63903	64647	64574	64773	64661	65107	65374	65363	65638	65769
02	All Scheduled Imports	767	923	809	928	813	884	786	1033	791	1121
03	All Scheduled Exports	3125	1663	2234	1571	2173	1421	1993	1425	1791	1427
04	Total Resources (01+02−03)	61545	63907	63149	64130	63301	64570	64167	64971	64638	65463
	Demand in MW										
05	Peak Hour Demand	36751	47683	37536	48753	38392	50018	39217	51070	39894	51989
06	Interruptible Demand	1088	1085	1023	1021	1025	1018	1047	1071	1038	1032
07	Demand Requirements (05−06)	35663	46598	36513	47732	37367	49000	38170	49999	38856	50957
	Margin in MW										
08	Margin (04−07)	25882	17309	26636	16398	25934	15570	25997	14972	25782	14506

		1992		1993		1994		1995		1996	
	Resources in MW	*Summer*	*Winter*	*Summer*	*Winter*	*Summer*	*Winter*	*Summer*	*Winter*	*Summer*	*Winter*
01	Net Dependable Capability	66065	65784	66108	65824	66173	66278	66539	66638	66964	66988
02	All Scheduled Imports	806	1126	814	1131	819	1136	854	1172	902	1209
03	All Scheduled Exports	1992	1629	2093	1730	2094	1601	1965	1503	1886	1506
04	Total Resources (01+02−03)	64879	65281	64829	65225	64898	65813	65428	66307	65980	66691
	Demand in MW										
05	Peak Hour Demand	40630	52939	41308	53906	42287	55042	43142	56048	43790	55828
06	Interruptible Demand	1026	1017	1024	1025	1042	1047	1055	1051	1040	1025
07	Demand Requirements (05−06)	39604	51922	40284	52881	41245	53995	42087	54997	42750	54803
	Margin in MW										
08	Margin (04−07)	25275	13359	24545	12344	23653	11818	23341	11310	23230	11888

Appendix D

A FERC Licensing Order

When the Federal Energy Regulatory Commission approves a hydroelectric proposal, it issues an order granting a license (or exemption) for the project. A licensing order includes detailed terms and conditions governing the construction and operation of the project, some of which are intended to mitigate project impacts on natural and recreational resources. The example below is an order granting a license for the Baldwin project on the Connecticut River in New Hampshire (42 FERC ¶ 62,007).

Robert W. Shaw, Project No. 7962–001
Order Issuing License (Major Project—5MW or Less)
(Issued January 11, 1988)
Fred E. Springer, Acting Director, Office of Hydropower Licensing

Robert W. Shaw has filed a license application under Part I of the Federal Power Act (Act) to construct, operate, and maintain the Baldwin Project, located in Coos County, New Hampshire, on the Connecticut River, a navigable waterway of the United States.

Notice of the application has been published. The motions to intervene that have been granted and the comments filed by agencies and individuals have been fully considered in determining whether to issue this license, as discussed below.

Recommendations of Federal and State Fish and Wildlife Agencies

Section 10(j) of Act, as amended by the Electric Consumers Protection Act of 1986 (ECPA), Public Law 99–495, requires the Commission to include license conditions, based on recommendations of federal and state fish and wildlife agencies, for the protection, mitigation, and enhancement of fish and wildlife. In the environmental assessment for the Baldwin Hydroelectric Project, the staff addresses the concerns of the federal and state fish and wildlife agencies, and makes recommendations consistent with those of the agencies.

Comprehensive Plans

Section 10(a)(2) of the Act, as amended by ECPA, requires the Commission to consider the extent to which a project is consistent with comprehensive plans (where they exist) for improving, developing, or conserving a waterway or waterways affected by the project. The plan must be prepared by an agency established pursuant to federal law that has the authority to prepare such a plan or by the state in which the facility is or will be located. The Commission considers plans to be within the scope of section 10(a)(2) only if such plans reflect the preparers' own balancing of the competing uses of a waterway, based on their data and on applicable policy considerations (i.e., if the preparer considers and balances all relevant public use considerations). With regard to plans prepared at the state level, such plans are within the scope of section 10(a)(2), only if they are prepared and adopted pursuant to a specific act of the state legislature and developed, implemented, and managed by an appropriate state agency.[1]

The Commission has concluded that comprehensive planning under section 10(a)(2)(A), like comprehensive planning under section 10(a)(1), should take into account all existing and potential uses of a waterway relevant to the public interest, including navigation, power development, energy conservation, fish and wildlife protection and enhancement, recreational opportunities, irrigation, flood control, water supply, and other aspects of environmental quality. In order that the Commission may fully understand or independently confirm the content and conclusions of a comprehensive plan, it provided general guidelines for developing such plans that should contain the following: (1) a description of the waterway(s) that are subject to the plan, including pertinent maps; (2) a description of the significant resources of the waterway(s); (3) a description of the various existing and planned uses for these resources; and (4) a discussion of goals, objectives, and recommendations for improving, developing, or conserving the waterway(s) in relation to these resources. The more closely a plan conforms to these guidelines, the more weight it will have on the Commission's decisions. The Commission, however, will consider plans that do not meet the criteria for comprehensive plans, as it considers all relevant studies and recommendations in its public interest analysis pursuant to section 10(a)(1) to the extent that the documentation supports the plan.[2]

The staff identified no comprehensive plans of the types referred to in section 10(a)(2) of the Act relevant to this project. As part of a broad public interest examination under Section 10(a)(1) of the Act, the staff reviewed four resource plans[3] that address various aspects of waterway management in relation to the proposed project. No conflicts were found.

Based on a review of agency and public comments filed in this proceeding, and on the staff's independent analysis, the Baldwin Hydroelectric Project is best adapted to a comprehensive plan for the Connecticut River, taking into consideration the beneficial public uses described in section 10(a)(2) of the Act.

Summary of Findings

An Environmental Assessment (EA) was issued for this project. Background information, analysis of impacts, support for related license articles, and the basis for a finding of no significant impact on the environment are contained in the EA attached to this order. Issuance of this license is not a major federal action significantly affecting the quality of the human environment.

The design of this project is consistent with the engineering standards governing dam safety. The project will be safe if constructed, operated, and maintained in accordance with the requirements of this license. Analysis of related issues is provided in the Safety and Design Assessment attached to this order.

The Director, Office of Hydropower Licensing, concludes that the project would not conflict with any planned or authorized development, and would be best adapted to comprehensive development of the waterway for beneficial public uses.

The Director orders:

(A) This license is issued to Robert W. Shaw (licensee), for a period of 50 years, effective the first day of the month in which this order is issued, to construct, operate, and maintain the Baldwin Project. This license is subject to the terms and conditions of the Act, which is incorporated by reference as part of this license, and subject to the regulations the Commission issues under the provision of the Act.

(B) The project consists of:

(1) All lands, to the extent of the licensee's interests in those lands, enclosed by the project boundary shown by Exhibit G:

Exhibit	*FERC No. 7962-*	*Showing*
G	8	Detail Map

(2) Project works consisting of: (a) the dam, 158 feet long, including an 8-foot-high, 66-foot-long spillway surmounted by 2-foot-high flashboards, and piers and gate structures having a combined length of 92 feet and a maximum height of 18 feet impounding a reservoir at elevation 1,299 feet having a surface area of 2 acres and a gross storage capacity of 16 acre-feet; (b) a reinforced concrete canal intake structure integral with the dam; (c) a 4,000-foot-long canal; (d) a penstock intake structure at the end of the canal; (e) a 10-foot-diameter, 300-foot-long penstock; (f) a reinforced concrete powerhouse containing a single 4,000 kW turbine/generator with a 600 cfs hydraulic capacity; (g) a 400-foot-long tailrace with water surface elevation 1,197 feet; (h) the 4.16-

kV generator leads; (i) the 4.16/34.5-kV, 5.5-MVA step-up transformer; (j) a 3,300-foot-long, 34.5-kV transmission line; and (k) appurtenant facilities.

The project works generally described above are more specifically shown and described by those portions of Exhibits A and F recommended for approval in the attached Safety and Design Assessment.

(3) All of the structures, fixtures, equipment or facilities used to operate or maintain the project and located within the project boundary, all portable property that may be employed in connection with the project and located within or outside the project boundary, and all riparian or other rights that are necessary or appropriate in the operation or maintenance of the project.

(C) The Exhibit G described above and those sections of Exhibits A and F recommended for approval in the attached Safety and Design Assessment are approved and made part of the license.

(D) This license is subject to the articles set forth in Form L-4 (October 1975) [reported at 54 FPC 1824], entitled "Terms and Conditions of License for Unconstructed Major Project Affecting Navigable Waters of the United States." The license is also subject to the following additional articles:

Article 201. The licensee shall pay the United States the following annual charge, effective the first day of the month in which this license is issued:

> For the purpose of reimbursing the United States for the cost of administration of Part I of the Act, a reasonable amount as determined in accordance with the provisions of the Commission's regulations in effect from time to time. The authorized installed capacity for that purpose is 5,400 horsepower.

Article 202. Pursuant to Section 10(d) of the Act, after the first 20 years of operation of the project under license, a specified reasonable rate of return upon the net investment in the project shall be used for determining surplus earnings of the project for the establishment and maintenance of amortization reserves. One-half of the project surplus earnings, if any, accumulated after the first 20 years of operation under the license, in excess of the specified rate of return per annum on the net investment, shall be set aside in a project amortization reserve account at the end of each fiscal year. To the extent that there is a deficiency of project earnings below the specified rate of return per annum for any fiscal year after the first 20 years of operation under the license, the amount of that deficiency shall be deducted from the amount of any surplus earnings subsequently accumulated, until absorbed. One-half of the remaining surplus earnings, if any, cumulatively computed, shall be set aside in the project amortization reserve account. The amounts established in the project amortization reserve account shall be maintained until further order of the Commission.

The annual specified reasonable rate of return shall be the sum of the annual weighted costs of long-term debt, preferred stock, and common equity, as defined below. The annual weighted cost for each component of the rate of return shall be calculated based on an average of 13 monthly balances of amounts properly includable in the licensee's long-term debt and proprietary capital accounts as listed in the Commission's Uniform System of Accounts. The cost rates for long-term debt and preferred stock shall be their respective weighted average costs for the year, and the cost of common equity shall be the interest rate on 10-year government bonds (reported as the Treasury Department's 10-year constant maturity series) computed on the monthly average for the year in question plus four percentage points (400 basis points).

Article 301. The licensee shall commence construction of project works within two years from the issuance date of the license and shall complete construction of the project within four years from the issuance date of the license.

Article 302. The licensee shall file, for approval by Commission, revised Exhibit F drawings showing the final design of project structures. The revised Exhibit F drawings shall be accompanied by a supporting design report and the licensee shall not commence construction of any project structure until the corresponding revised Exhibit F drawing has been approved.

Article 303. The licensee shall at least 60 days prior to the start of construction, submit one copy to the Commission's Regional Director and two copies to the Director, Division of Inspections, of the final contract drawings and specifications for pertinent features of the project, such as water retention structures, powerhouse, and water conveyance structures. The Director, Division of Inspections, may require changes in the plans and specifications to assure a safe and adequate project.

Article 304. The licensee shall review and approve the design of contractor-designed cofferdams and deep excavations prior to the start of construction and shall ensure that construction of cofferdams and deep excavations is consistent with the approved design. At least 30 days prior to the start of construction of the cofferdam, the licensee shall submit to the Commission's Regional Director and the Director, Division of Inspections, one copy each of the approved cofferdam construction drawings and specifications and the letter(s) of approval.

Article 305. The licensee shall, within 90 days of completion of construction, file for approval by the Commission, revised Exhibits A, F, and G to describe and show the project as built.

Article 401. The licensee, after consultation with the U.S. Fish and Wildlife Service and the New Hampshire Fish and Game Department, and within 1 year of the date of issuance of this license, shall file with the Commission, a comprehensive plan to control erosion, dust, and bank stability and to minimize the quantity of sediment or other potential water pollutants resulting from project construction, spoil disposal, and project operation, for the protection of aquatic resources. The plan shall include the reservoir shoreline and tailrace channel. The plan shall also include descriptions and functional design drawings of control measures, topographic map locations of control measures, an implementation schedule, monitoring and maintenance programs for project construction and operation, and provisions for periodic review of the plan and for making any necessary revisions to the plan. The licensee shall include in the filing documentation of agency consultation on the plan, and copies of agency comments or recommendations.

If the licensee disagrees with any agency recommendations, the licensee shall provide a discussion of the reasons for disagreeing, based on actual-site geological, soil, and groundwater conditions. The Commission reserves the right to require changes to the plan. Unless the Director of the Office of Hydropower Licensing directs otherwise, the licensee may commence project-related land-clearing, land-disturbing, and spoil-producing activities at the project, 60 days after filing this plan.

Article 402. The licensee shall discharge from the Baldwin Dam Project a continuous minimum flow of 60 cubic feet per second, as measured immediately downstream from the Baldwin dam, or the inflow to the reservoir, whichever is less, for the protection of fish and wildlife resources in the Connecticut River. This flow may be temporarily modified if required by operating emergencies beyond the control of the licensee and for short periods upon mutual agreement between the licensee and the New Hampshire Fish and Game Department.

Article 403. The licensee shall consult with the New Hampshire Fish and Game Department and the U.S. Fish and Wildlife Service in developing a plan to ensure the continual release of the minimum flow required by article 402. The plan must include the design and hydraulic sizing calculations of the device and the technique used to pass the required minimum flow at Baldwin Dam. Further, the licensee shall file the plan for Commission approval no later than 60 days before beginning project construction, and shall include comments on the plan from the consulted agencies. The Commission reserves the right to require modifications to the plan. Further, before beginning project operation, the licensee must be capable of discharging the required minimum flow from the dam in accordance with the plan approved by the Commission.

Article 404. The licensee shall operate the Baldwin Project in an instantaneous run-of-river mode for the protection of fish and wildlife resources in the Connecticut River. The licensee, in operating the project in an instantaneous run-of-river mode, shall at all times act to minimize the fluctuation of the reservoir surface elevation by maintaining a sufficient discharge from the project so that flow in the Connecticut River, as measured immediately downstream from the project tailrace, approximates the instantaneous sum of the inflow to the project reservoir. Instantaneous run-of-river operation may be temporarily modified if required by operation emergencies beyond control of the licensee and for short periods upon mutual agreement between the licensee and the New Hampshire Fish and Game Department.

Article 405. The licensee must construct, operate, and maintain, or arrange for the construction, operation, and maintenance of downstream fish passage facilities for the Baldwin Project. The licensee shall consult with the New Hampshire Fish and Game Department and the U.S. Fish and Wildlife Service and shall develop a plan to minimize the impacts of project operation on outmigrating salmonids in the Connecticut River. The plan shall include functional design drawings of downstream fish passage facilities, including an appropriate screen on the intake structure and a schedule for construction, operation, and maintenance of the facilities. Further, the licensee shall file the plan for Commission approval no later than 60 days before beginning project construction, and shall include comments on the plan from the consulted agencies. The Commission reserves the right to require modifications to the plan. Before beginning project operation, the licensee must complete construction of downstream fish passage facilities and of the intake screening, in accordance with the plan approved by the Commission.

Article 406. The licensee, after consultation with the U.S. Fish and Wildlife Service (FWS) and the New Hampshire Fish and Game Department (FGD), and within 1 year after the date of issuance of this license, shall develop and file with the Commission, a plan to revegetate all areas disturbed during project construction with plant species beneficial to wildlife. The plan, at a minimum, shall include the following: (1) a description of the plant species and the densities to be used; (2) a monitoring program to evaluate the effectiveness of the revegetation; (3) a description of the procedures to be followed if monitoring indicates that revegetation is not successful; and (4) an implementation schedule. The licensee shall include in the filing comments from FWS and FGD on the plan. The Commission reserves the right to require changes to the plan.

Article 407. The licensee, before starting any ground-disturbing or land-clearing activities within the project boundaries, other than those specifically authorized in this license, shall consult the State Historic Preservation Officer (SHPO) about the need for a cultural resources survey and for salvage work. The licensee shall file with the Commission documentation of the nature and extent of consultation, including a cultural resources management plan and a schedule to conduct the necessary investigation, together with a copy of a letter from the SHPO commenting on the plan and schedule, sixty days before starting any such ground-disturbing or land-clearing activities. The licensee shall make funds available in a reasonable amount for the required work. If the licensee discovers any previously unidentified archeological or historic sites during the course of constructing or developing project works or other facilities at the project, the licensee shall stop all construction and development activities in the vicinity of the sites and shall consult a qualified cultural resources specialist and the SHPO about the eligibility of the sites for listing in the *National Register of Historic Places* and about any measures needed to avoid the sites or to mitigate effects on the sites. If the licensee and the SHPO cannot agree on the amount of money to be spent for project specific archeological and historical purposes, the Commission reserves the right to require the licensee to conduct the necessary work at the licensee's own expense.

Article 408. (a) In accordance with the provisions of this article, the licensee shall have the authority to grant permission for certain types of use and occupancy of project lands and waters and to convey certain interests in project lands and waters for certain other types of use and occupancy, without prior Commission approval. The licensee may exercise the authority only if the proposed use and occupancy is consistent with the purposes of protecting and enhancing the scenic, recreational, and other environmental values of the project. For those purposes, the licensee shall also have continuing responsibility to supervise and control the uses and occupancies for which it grants permission, and to monitor the use of, and ensure compliance with the covenants of the instrument of conveyance for, any interests that it has conveyed, under this article. If a permitted use and occupancy violates any condition of this article or any other condition imposed by the licensee for protection and enhancement of the project's scenic, recreational, or other environmental values, or if a covenant of a conveyance made under the authority of this article is violated, the licensee shall take any lawful action necessary to correct the violation. For a permitted use or occupancy, that action includes, if necessary, cancelling the permission to use and occupy the project lands and waters and requiring the removal of any non-complying structures and facilities.

(b) The types of use and occupancy of project lands and waters for which the licensee may grant permission without prior Commission approval are: (1) landscape plantings; (2) non-commercial piers, landings, boat docks, or similar structures and facilities that can accommodate no more than 10 watercraft at a time and where said facility is intended to serve single-family type dwellings; and (3) embankments, bulkheads, retaining walls, or similar structures for erosion control to protect the existing shoreline. To the extent feasible and desirable to protect and enhance the project's scenic, recreational, and other environmental values, the licensee shall require multiple use and occupancy of facilities for access to project lands or waters. The licensee shall also ensure, to the satisfaction of the Commission's authorized representative, that the uses and occupancies for which it grants permission are maintained in good repair and comply with applicable state and local health and safety requirements. Before granting permission for construction of bulkheads or retaining walls, the licensee shall: (1) inspect the site of the proposed construction, (2) consider whether the planting of vegetation or the use of riprap would be adequate to control erosion at the site, and (3) determine that the proposed construction is needed and would not change the basic contour of the reservoir shoreline. To implement this paragraph (b), the licensee may, among other things, establish a program for issuing permits for the specified types of use and occupancy of project lands and waters, which may be subject to the payment of a reasonable fee to cover the licensee's costs of administering the permit program. The Commission reserves the right to require the licensee to file a description of its standards, guidelines, and procedures for implementing this paragraph (b) and to require modification of those standards, guidelines, or procedures.

(c) The licensee may convey easements or rights-of-way across, or leases of, project lands for: (1) replacement, expansion, realignment, or maintenance of bridges and roads for which all necessary state and federal approvals have been obtained; (2) storm drains and water mains; (3) sewers that do not discharge into project waters; (4) minor access roads; (5) telephone, gas, and electric utility distribution lines; (6) non-project overhead electric transmission lines that do not require erection of support structures within the project boundary; (7) submarine, overhead, or underground major telephone distribution cables or major electric distribution lines (69–kV or less); and (8) water intake or pumping facilities that do not extract more than one million gallons per day from a project reservoir. No later than January 31 of each year, the licensee shall file three copies of a report briefly describing for each conveyance made under this paragraph (c) during the prior calendar year, the type of interest conveyed, the location of the lands subject to the conveyance, and the nature of the use for which the interest was conveyed.

(d) The licensee may convey fee title to, easements or rights-of-way across, or leases of project lands for: (1) construction of new bridges or roads for which all necessary state and federal approvals have been obtained; (2) sewer or effluent lines that discharge into project waters, for which all necessary federal and state water quality certificates or permits have been obtained; (3) other pipelines that cross project lands or waters but do not discharge into project waters; (4) non-project overhead electric transmission lines that require erection of support structures within the project boundary, for which all necessary federal and state approvals have been obtained; (5) private or public marinas that can accommodate no more than 10 watercraft at a time and are located at least one-half mile from any other private or public marina; (6) recreational development consistent with an approved Exhibit R or approved report on recreational resources of an Exhibit E; and (7) other uses, if: (i) the amount of land conveyed for a particular use is five acres or less; (ii) all of the land conveyed is located at least 75 feet, measured horizontally, from the edge of the project reservoir at normal maximum surface elevation; and (iii) no more than 50 total acres of project lands for each project development are conveyed under this clause (d)(7) in any calendar year. At least 45 days before conveying any interest in project lands under this paragraph (d), the licensee must submit a letter to the Director, Office of Hydropower Licensing, stating its intent to convey the interest and briefly describing the type of interest and location of the lands to be conveyed (a marked Exhibit G map may be used), the nature of the proposed use, the identity of any federal or state agency official consulted, and any federal or state approvals required for the proposed use. Unless the Director, within 45 days from the filing date, requires the licensee to file an application for prior approval, the licensee may convey the intended interest at the end of that period.

(e) The following additional conditions apply to any intended conveyance under paragraph (c) or (d) of this article:

> (1) Before conveying the interest, the licensee shall consult with federal and state fish and wildlife or recreation agencies, as appropriate, and the State Historic Preservation Officer.
>
> (2) Before conveying the interest, the licensee shall determine that the proposed use of the lands to be conveyed is not inconsistent with any approved Exhibit R or approved report on recreational resources of an Exhibit E; or, if the project does not have an approved Exhibit R or approved report on recreational resources, that the lands to be conveyed do not have recreational value.
>
> (3) The instrument of conveyance must include covenants running with the land adequate to ensure that: (i) the use of the lands conveyed shall not endanger health, create a nuisance, or otherwise be incompatible with overall project recreational use; and (ii) the grantee shall take all reasonable precautions to ensure that the construction, operation, and maintenance of structures or facilities on the conveyed lands will occur in a manner that will protect the scenic, recreational, and environmental values of the project.
>
> (4) The Commission reserves the right to require the licensee to take reasonable remedial action to correct any violation of the terms and conditions of this article, for the protection and enhancement of the project's scenic, recreational, and other environmental values.

(f) The conveyance of an interest in project lands under this article does not in itself change the project boundaries. The project boundaries may be changed to exclude land conveyed under this article only upon approval of revised Exhibit G drawings (project boundary maps) reflecting exclusion of that land. Lands conveyed under this article will be excluded from the project only upon a determination that the lands are not necessary for project purposes, such as operation and maintenance, flowage, recreation, public access, protection of environmental resources, and shoreline control, including shoreline aesthetic values. Absent extraordinary circumstances, proposals to exclude lands conveyed under this article from the project shall be consolidated for consideration when revised Exhibit G drawings would be filed for approval for other purposes.

(g) The authority granted to the licensee under this article shall not apply to any part of the public lands and reservations of the United States included within the project boundary.

(E) The licensee shall serve copies of any Commission filing required by this order on any entity specified in this order to be consulted on matters related to that filing. Proof of service on these entities must accompany the filing with the Commission.

(F) This order is issued under authority delegated to the Director and is final unless appealed under Rule 1902 to the Commission by any party within 30 days from the issuance date of this order. Filing an appeal does not stay the effective date of this order or any date specified in this order. The licensee's failure to appeal this order shall constitute acceptance of the license.

—Footnotes—

[1] *See Fieldcrest Mills, Inc.*, 37 FERC ¶ 61,264 (1986).

[2] *See* Commission Order No. 481, issued October 20, 1987 [*FERC Statutes and Regulations* ¶ 30,773].

[3] New Hampshire Water Resources Management Plan, 1984, New Hampshire Water Resources Board; Fish and Wildlife Plan for New Hampshire, 1983, New Hampshire Fish and Game Department; Wild, Scenic, and Recreational Rivers for New Hampshire, 1977, New Hampshire Office of State Planning; Waterfowl and Their Management in New Hampshire, 1975, New Hampshire Fish and Game Department.

Appendix E

FERC's NEPA Regulations

The National Environmental Policy Act of 1969 (NEPA) requires all federal agencies to prepare written statements on the environmental impacts of federal actions that may significantly affect the human environment. Every federal agency has promulgated detailed regulations explaining its standards and procedures for preparing environmental impact statements and environmental assessments under NEPA. This appendix reprints FERC's NEPA regulations from the Federal Register [*52 Fed. Reg. 47910 (17 December 1987)*].

PART 380—REGULATIONS IMPLEMENTING THE NATIONAL ENVIRONMENTAL POLICY ACT

Sec.
380.1 Purpose.
380.2 Definitions and terminology.
380.3 Environmental information to be supplied by an applicant.
380.4 Projects or actions categorically excluded.
380.5 Actions that require an environmental assessment.
380.6 Actions that require an environmental impact statement.
380.7 Format of an environmental impact statement.
380.8 Preparation of environmental documents.
380.9 Public availability of NEPA documents and public notice of NEPA related hearings and public meetings.

380.10 Participation in Commission proceedings.
380.11 Environmental decisionmaking.

Authority: National Environmental Policy Act of 1969, 42 U.S.C. 4321–4370a (1982); Department of Energy Organization Act, 42 U.S.C. 7101–7352 (1982); E.O. No. 12009, 3 CFR 1978 Comp., p. 142.

§ 380.1 Purpose.

The regulations in this part implement the Federal Energy Regulatory Commission's procedures under the National Environmental Policy Act of 1969. These regulations supplement the regulations of the Council on Environmental Quality, 40 CFR Parts 1500 through 1508 (1986). The Commission will comply with the regulations of the Council on Environmental Quality except where those regulations are inconsistent with the statutory requirements of the Commission.

§ 380.2 Definitions and terminology.

For purposes of this part—

(a) "Categorical exclusion" means a category of actions described in § 380.4, which do not individually or cumulatively have a significant effect on the human environment and which the Commission has found to have no such effect and for which, therefore, neither an environmental assessment nor an environmental impact statement is required. The Commission may decide to prepare environmental assessments for the reasons stated in § 380.4(b).

(b) "Commission" means the Federal Energy Regulatory Commission.

(c) "Council" means the Council on Environmental Quality.

(d) "Environmental assessment" means a concise public document for which the Commission is responsible that serves to:

(1) Briefly provide sufficient evidence and analysis for determining whether to prepare an environmental impact statement or a finding of no significant impact.

(2) Aid the Commission's compliance with NEPA when no environmental impact statement is necessary.

(3) Facilitate preparation of a statement when one is necessary. Environmental assessments must include brief discussions of the need for the proposal, of alternatives as required by section 102(2)(E) of NEPA, of the environmental impacts of the proposed action and alternatives, and a listing of agencies and persons consulted.

(e) "Environmental impact statement" (EIS) means a detailed written statement as required by section 102(2)(C) of NEPA. DEIS means a draft EIS and FEIS means a final EIS.

(f) "Environmental report" or ER means that part of an application submitted to the Commission by an applicant for authorization of a proposed action which includes information concerning the environment, the applicant's analysis of the environmental impact of the action, or alternatives to the action required by this or other applicable statutes or regulations.

(g) "Finding of no significant impact" (FONSI) means a document by the Commission briefly presenting the reason why an action, not otherwise excluded by § 380.4, will not have a significant effect on the human environment and for which an environmental impact statement therefore will not be prepared. It must include the environmental assessment or a summary of it and must note other environmental documents related to it. If the assessment is included, the FONSI need not repeat any of the discussion in the assessment but may incorporate it by reference.

§ 380.3 Environmental information to be supplied by an applicant.

(a) An applicant must submit information as follows:

(1) For any proposed action identified in §§ 380.5 and 380.6, an environmental report with the proposal as prescribed in paragraph (c) of this section.

(2) For any proposal not identified in paragraph (a)(1) of this section, any environmental information that the Commission may determine is necessary for compliance with these regulations, the regulations of the Council, NEPA and other Federal laws such as the Endangered Species Act, the National Historic Preservation Act or the Coastal Zone Management Act.

(b) An applicant must also:

(1) Provide all necessary or relevant information to the Commission;

(2) Conduct any studies that the Commission staff considers necessary or relevant to determine the impact of the proposal on the human environment and natural resources;

(3) Consult with appropriate Federal, regional, State, and local agencies during the planning stages of the proposed action to ensure that all potential environmental impacts are identified. (The specific requirements for consultation on hydropower projects are contained in § 4.38 of this chapter and in section 4(a) of the Electric Consumers Protection Act, Pub. L. No. 99-495, 100 Stat. 1243, 1246 (1986));

(4) Submit applications for all Federal and State approvals as early as possible in the planning process; and

(5) Notify the Commission staff of all other Federal actions required for completion of the proposed action so that the staff may coordinate with other interested Federal agencies.

(c) *Content of an applicant's environmental report for specific proposals*—(1) *Hydropower projects.* The information required for specific project applications under Part 4 of this chapter.

(2) *Natural gas projects.* (i) For any application filed under the Natural Gas Act for any proposed action identified in §§ 380.5 or 380.6, except for prior notice filings under § 157.208, as described in § 380.5(b), the information identified in Appendix A of this part.

(ii) For prior notice filings under § 157.208, the report described by § 157.208(c)(11) of this chapter.

§ 380.4 Projects or actions categorically excluded.

(a) *General rule.* Except as stated in paragraph (b) of this section, neither an environmental assessment nor an environmental impact statement will be prepared for the following projects or actions:

(1) Procedural, ministerial, or internal administrative and management actions, programs, or decisions, including procurement, contracting, personnel actions, correction or clarification of filings or orders, and acceptance, rejection and dismissal of filings;

(2)(i) Reports or recommendations on legislation not initiated by the Commission, and

(ii) Proposals for legislation and promulgation of rules that are clarifying, corrective, or procedural, or that do not substantially change the effect of legislation or regulations being amended;

(3) Compliance and review actions, including investigations (jurisdictional or otherwise), conferences, hearings, notices of probable violation, show cause orders, and adjustments under section 502(c) of the Natural Gas Policy Act of 1978 (NGPA);

(4) Review of grants or denials by the Department of Energy (DOE) of any adjustment request, and review of contested remedial orders issued by DOE;

(5) Information gathering, analysis, and dissemination;

(6) Conceptual or feasibility studies;

(7) Actions concerning the reservation and classification of United States lands as water power sites and other actions under section 24 of the Federal Power Act;

(8) Transfers of water power project licenses and transfers of exemptions under Part I of the Federal Power Act and Part 9 of this chapter;

(9) Issuance of preliminary permits for water power projects under Part I of the Federal Power Act and Part 4 of this chapter;

(10) Withdrawals of applications for certificates under the Natural Gas Act, or for water power project preliminary permits, exemptions, or licenses under Part I of the Federal Power Act and Part 4 of this chapter;

(11) Actions concerning annual charges or headwater benefits, charges for water power projects under Parts 11 and 13 of this chapter and establishment of fees to be paid by an applicant for a license or exemption required to meet the terms and conditions of section 30(c) of the Federal Power Act;

(12) Approval for water power projects under Part I of the Federal Power Act, of "as built" or revised drawings or exhibits that propose no changes to project works or operations or that reflect changes that have previously been approved or required by the Commission;

(13) Surrender and amendment of preliminary permits, and surrender of water power licenses and exemptions where no project works exist or ground disturbing activity has occurred and amendments to water power licenses and exemptions that do not require ground disturbing activity or changes to project works or operation;

(14) Exemptions for small conduit hydroelectric facilities as defined in

§ 4.30(b)(26); of this chapter under Part I of the Federal Power Act and Part 4 of this chapter;

(15) Electric rate filings submitted by public utilities, establishment of just and reasonable rates, and confirmation, approval, and disapproval of rate filings submitted by Federal power marketing agencies under sections 205 and 206 of the Federal Power Act;

(16) Approval of actions under sections 4(b), 203, 204, 301, 304, and 305 of the Federal Power Act relating to issuance and purchase of securities, acquisition or disposition of property, merger, interlocking directorates, jurisdictional determinations and accounting orders;

(17) Approval of electrical interconnections and wheeling under sections 202(b), 210, 211, and 212 of the Federal Power Act, that would not entail:

(i) Construction of a new substation or expansion of the boundaries of an existing substation;

(ii) Construction of any transmission line that operates at more than 115 kilovolts (KV) and occupies more than ten miles of an existing right-of-way; or

(iii) Construction of any transmission line more than one mile long if located on a new right-of-way;

(18) Approval of changes in land rights for water power projects under Part I of the Federal Power Act and Part 4 of this chapter, if no construction or change in land use is either proposed or known by the Commission to be contemplated for the land affected;

(19) Approval of proposals under Part I of the Federal Power Act and Part 4 of this chapter to authorize use of water power project lands or waters for gas or electric utility distribution lines, radial (sub-transmission) lines, communications lines and cables, storm drains, sewer lines not discharging into project waters, water mains, piers, landings, boat docks, or similar structures and facilities, landscaping or embankments, bulkheads, retaining walls, or similar shoreline erosion control structures;

(20) Action on applications for exemption under section 1(c) of the Natural Gas Act;

(21) Approvals of blanket certificate applications and prior notice filings under § 157.204 and §§ 157.209 through 157.218 of this chapter;

(22) Approvals of blanket certificate applications under §§ 284.221 through 284.224 of this chapter;

(23) Producers' applications for the sale of gas filed under §§ 157.23 through 157.29 of this chapter;

(24) Approval under section 7 of the Natural Gas Act of taps, meters, and regulating facilities located completely within an existing natural gas pipeline right-of-way or compressor station if company records show the land use of the vicinity has not changed since the original facilities were installed, and no significant nonjurisdictional facilities would be constructed in association with construction of the interconnection facilities;

(25) Review of natural gas rate filings, including any curtailment plans other than those specified in § 380.5(b)(5), and establishment of rates for trans-

portation and sale of natural gas under sections 4 and 5 of the Natural Gas Act and sections 311 and 401 through 404 of the Natural Gas Policy Act of 1978;

(26) Review of approval of oil pipeline rate filings under Parts 340 and 341 of this chapter;

(27) Sale, exchange, and transportation of natural gas under sections 4, 5 and 7 of the Natural Gas Act that require no construction of facilities;

(28) Abandonment in place of a minor natural gas pipeline (short segments of buried pipe of 6-inch inside diameter or less), or abandonment by removal of minor surface facilities such as metering stations, valves, and tops under section 7 of the Natural Gas Act so long as appropriate erosion control and site restoration take place;

(29) Abandonment of service under any gas supply contract pursuant to section 7 of the Natural Gas Act;

(30) Approval of filing made in compliance with the requirements of a certificate for a natural gas project under section 7 of the Natural Gas Act or a preliminary permit, exemption, license, or license amendment order for a water power project under Part I of the Federal Power Act;

(b) *Exceptions to categorical exclusions.* (1) In accordance with 40 CFR 1508.4, the Commission and its staff will independently evaluate environmental information supplied in an application and in comments by the public. Where circumstances indicate that an action may be a major Federal action significantly affecting the quality of the human environment, the Commission:

(i) May require an environmental report or other additional environmental information, and

(ii) Will prepare an environmental assessment or an environmental impact statement.

(2) Such circumstances may exist when the action may have an effect on one of the following:

(i) Indian lands;

(ii) Wilderness areas;

(iii) Wild and scenic rivers;

(iv) Wetlands;

(v) Units of the National Park System, National Refuges, or National Fish Hatcheries;

(vi) Anadromous fish or endangered species; or

(vii) Where the environmental effects are uncertain.

However, the existence of one or more of the above will not automatically require the submission of an environmental report or the preparation of an environmental assessment or an environmental impact statement.

§ 380.5 Actions that require an environmental assessment.

(a) An environmental assessment will normally be prepared first for the actions identified in this section. Depending on the outcome of the environmental assessment, the Commission may or may not prepare an environmental impact statement. However, depending on the location or scope of the proposed action, or the resources affected, the Commission may in specific circumstances proceed directly to prepare an environmental impact statement.

(b) The projects subject to an environmental assessment are as follows:

(1) Except as identified in §§ 380.4, 380.6 and 2.55 of this chapter, authorization under section 7 of the Natural Gas Act for the construction, replacement, or abandonment of compression, processing, or interconnecting facilities, onshore and offshore pipelines, metering facilities, LNG peak-shaving facilities, or other facilities necessary for the sale, exchange, storage, or transportation of natural gas;

(2) Prior notice filings under § 157.208 of this chapter for the rearrangement of any facility specified in §§ 157.202 (b)(3) and (6) of this chapter or the acquisition, construction, or operation of any eligible facility as specified in §§ 157.202 (b)(2) and (3) of this chapter;

(3) Abandonment or reduction of natural gas service under section 7 of the Natural Gas Act unless excluded under §§ 380.4 (a)(21), (28) or (29);

(4) Except as identified in § 380.6, conversion of existing depleted oil or natural gas fields to underground storage fields under section 7 of the Natural Gas Act;

(5) New natural gas curtailment plans, or any amendment to an existing curtailment plan under section 4 of the Natural Gas Act and sections 401 through 404 of the Natural Gas Policy Act of 1978 that has a major effect on an entire pipeline system;

(6) Licenses under Part I of the Federal Power Act and Part 4 of this chapter for construction of any water power project—existing dam;

(7) Exemptions under section 405 of the Public Utility Regulatory Policies Act of 1978, as amended, and §§ 4.30(b)(27) and 4.101–4.106 of this chapter for small hydroelectric power projects of 5 MW or less;

(8) Licenses for additional project works at licensed projects under Part I of the Federal Power Act whether or not these are styled license amendments or original licenses;

(9) Licenses under Part I of the Federal Power Act and Part 4 of this chapter for transmission lines only;

(10) Applications for new licenses under section 15 of the Federal Power Act;

(11) Approval of electric interconnections and wheeling under sections 202(b), 210, 211, and 212 of the Federal Power Act, unless excluded under § 380.4(a)(17); and

(12) Regulations or proposals for legislation not excluded under § 380.4(a)(2).

(13) Surrender of water power licenses and exemptions where project works exist or ground disturbing activity has occurred and amendments to water power licenses and exemptions that require ground disturbing activity or changes to project works or operations.

§ 380.6 Actions that require an environmental impact statement.

(a) Except as provided in paragraph (b) of this section, an environmental impact statement will normally be prepared first for the following projects:

(1) Authorization under section 3 or 7 of the Natural Gas Act for construction and operation of jurisdictional liquefied natural gas import/export facilities used wholly or in part to liquefy, store, or regasify liquefied natural gas transported by water;

(2) Certificate applications under section 7 of the Natural Gas Act to develop an underground natural gas storage facility except where depleted oil or natural gas producing fields are used;

(3) Major pipeline construction projects under section 7 of the Natural Gas Act using right-of-way in which there is no existing natural gas pipeline; and

(4) Licenses under Part I of the Federal Power Act and Part 4 of this chapter for construction of any unconstructed water power project.

(b) If the Commission believes that a proposed action identified in paragraph (a) of this section may not be a major Federal action significantly affecting the quality of the human environment, an environmental assessment, rather than an environmental impact statement, will be prepared first. Depending on the outcome of the environmental assessment, an environmental impact statement may or may not be prepared.

(c) An environmental impact statement will not be required if an environmental assessment indicates that a proposal has adverse environmental affects and the proposal is not approved.

§ 380.7 Format of an environmental impact statement.

In addition to the requirements for an environmental impact statement prescribed in 40 CFR 1502.10 of the regulations of the Council, an environmental impact statement prepared by the Commission will include a section on the literature cited in the environmental impact statement and a staff conclusion section. The staff conclusion section will include summaries of:

(a) The significant environmental impacts of the proposed action;

(b) Any alternative to the proposed action that would have a less severe environmental impact or impacts and the action preferred by the staff;

(c) Any mitigation measures proposed by the applicant, as well as additional mitigation measures that might be more effective;

(d) Any significant environmental impacts of the proposed action that cannot be mitigated; and

(e) References to any pending, completed, or recommended studies that might provide baseline data or additional data on the proposed action.

§ 380.8 Preparation of environmental documents.

The preparation of environmental documents, as defined in § 1508.10 of the regulations of the Council, on hydroelectric projects, is the responsibility of the Commission's Office of Hydropower Licensing, 400 First Street NW, Washington, DC 20426, (202) 376–9171. The preparation of environmental documents on natural gas projects is the responsibility of the Commission's Office of Pipeline and Producer Regulation, (202) 357–8500, 825 North Capitol Street NE, Washington, DC 20426. Persons interested in status reports or information on environmental impact statements or other elements of the NEPA process, including the studies or other information the Commission may require on these projects, can contact these sections.

§ 380.9 Public availability of NEPA documents and public notice of NEPA related hearings and public meetings.

(a)(1) The Commission will comply with the requirements of 40 CFR 1506.6 of the regulations of the Council for public involvement in NEPA.

(2) If an action has effects of primarily local concern, the Commission may give additional notice in a Commission order.

(b) The Commission will make environmental impact statements, environmental assessments, the comments received, and any underlaying documents available to the public pursuant to the provisions of the Freedom of Information Act [5 U.S.C. 552 (1982)]. The exclusion in the Freedom of Information Act for interagency memoranda is not applicable where such memoranda transmit comments of Federal agencies on the environmental impact of the proposed action. Such materials will be made available to the public at the Commission's Public Reference Room at 825 North Capitol Street NE, Room 1000, Washington, DC 20426 at a fee and in the manner described in Part 388 of this chapter. A copy of an environmental impact statement or environmental assessment for hydroelectric projects may also be made available for inspection at the Commission's regional office for the region where the proposed action is located.

§ 380.10 Participation in Commission proceedings.

(a) *Intervention proceedings involving a party or parties*—(1) *Motion to intervene.* (i) In addition to submitting comments on the NEPA process and NEPA related documents, any person may file a motion to intervene in a Commission proceeding dealing with environmental issues under the terms of § 385.214 of this chapter. Any person who files a motion to intervene on the basis of a draft environmental impact statement will be deemed to have filed a timely motion, in accordance with § 385.214, as long as the motion is filed within the comment period for the draft environmental impact statement.

(ii) Any person that is granted intervention after petitioning becomes a party to the proceeding and accepts the record as developed by the parties as of the time that intervention is granted.

(2)(i) *Issues not set for trial-type hearing.* An intervenor who takes a position on any environmental issue that has not yet been set for hearing must file a timely motion with the Secretary containing an analysis of its position on such issue and specifying any differences with the position of Commission staff or an applicant upon which the intervenor wishes to be heard at a hearing.

(ii) *Issues set for trial-type hearing.* (A) Any intervenor that takes a position on an environmental issue set for hearing may offer evidence for the record in support of such position and otherwise participate in accordance with the Commission's Rules of Practice and Procedure. Any intervenor must specify any differences from the staff's and the applicant's positions.

(B) To be considered, any facts or opinions on an environmental issue set for hearing must be admitted into evidence and made part of the record of the proceeding.

(b) *Rulemaking proceedings.* Any person may file comments on any environmental issue in a rulemaking proceeding.

§ 380.11 Environmental decisionmaking.

(a) *Decision points.* For the actions which require an environmental assessment or environmental impact statement, environmental considerations will be addressed at appropriate major decision points.

(1) In proceedings involving a party or parties and not set for trial-type hearing, major decision points are the approval or denial of proposals by the Commission or its designees.

(2) In matters set for trial-type hearing, the major decision points are the initial decision of an administrative law judge or the decision of the Commission.

(3) In a rulemaking proceeding, the major decision points are the Notice of Proposed Rulemaking and the Final Rule.

(b) *Environmental documents as part of the record.* The Commission will include environmental assessments, findings of no significant impact, or environmental impact statements, and any supplements in the record of the proceeding.

(c) *Application denials.* Notwithstanding any provision in this Part, the Commission may dismiss or deny an application without performing an environmental impact statement or without undertaking environmental analysis.

Appendix F

FERC's Freedom of Information Regulations

The Freedom of Information Act (FOIA) gives the public a general right of access to documents held in agency files, subject to various narrowly drawn exceptions. You can obtain most of the documents you need upon request from the public information room at FERC, but on occasion it may be necessary to make a formal FOIA request to the Commission. The Commission's FOIA regulations, which are set forth below, explain the steps necessary to pursue a FOIA request [53 Fed. Reg. 1473 (20 January 1988)].

PART 388—INFORMATION AND REQUESTS

Sec.

388.101 Scope.
388.102 Notice of proceedings.
388.103 Notice and publication of decisions, rules, statements of policy, organization and operations.
388.104 Informal advice from Commission staff.
388.105 Procedures for press, television, radio, and photographic coverage.
388.106 Requests for Commission records available in the Public Reference Room.
388.107 Commission records exempt from public disclosure.
388.108 Requests for Commission records not available through the Public Reference Room (FOIA requests).
388.109 Fees for record requests.

388.110 Procedure for appeal of denial of requests for Commission records not publicly available or not available through the Public Reference Room and denial of requests for fee waiver or reduction.
388.111 Procedures in event of subpoena.
388.112 Requests for privileged treatment of documents submitted to the Commission.

Authority: Freedom of Information Act, 5 U.S.C. 552 (1982) as amended by Freedom of Information Reform Act of 1986; Administrative Procedure Act, 5 U.S.C. 551–557 (1982).

§ 388.101 Scope.

This part prescribes the rules governing public notice of proceedings, publication of decisions, requests for informal advice from Commission staff, procedures for press, television, radio and photographic coverage, requests for Commission records, requests for confidential treatment of documents submitted to the Commission, procedures for responding to subpoenas seeking documents or testimony from Commission employees or former employees, fees for various requests for documents, and requests for reduction or waiver of these fees.

§ 388.102 Notice of proceedings.

(a) Public sessions of the Commission for taking evidence or hearing argument; public conferences and hearings before a presiding officer; and public conferences or hearings in substantive rulemaking proceedings, will not be held except upon notice.

(b) Notice of applications, complaints, and petitions, is governed by Rule 2009 (notice) in Part 385 of this chapter. Notice of applications for certificates of public convenience and necessity under section 7 of the Natural Gas Act is governed by § 157.9 of this chapter (notice of application). Notice of public sessions and proceedings and of meetings of the Commission is governed by Rule 2009 (notice) in Part 385 of this Chapter. Notice of hearings and of initiation or pendency of rulemaking proceedings is governed by Rule 1903 (notice in rulemaking proceedings) in Part 385 of this chapter. Notice of application under Part I of the Federal Power Act for preliminary permits and licenses is governed by §§ 4.31 and 4.81 of this chapter (acceptance or rejection and contents). Notice of proposed alterations or surrenders of license under section 6 of the Federal Power Act may be given by filing and publication in the **Federal Register** as stated in Rule 1903 (notice in rulemaking proceedings) in Part 385 of this chapter, and where deemed desirable by the Commission, by local newspaper advertisement. Notice of rates charged and changes therein is governed by the filing requirements of subchapters B and E of this chapter (regulations under the Federal Power Act and regulations under the Natural Gas Act). Other notice required by statute, rule, regulation, or order, or deemed desirable, may be given by filing and publication in the **Federal**

Register as governed by Rule 1903 in Part 385 of this chapter (notice in rulemaking proceedings) or by service as governed by Rule 2010 (service) in Part 385 of this chapter.

§ 388.103 Notice and publication of decisions, rules, statements of policy, organization and operations.

Service of intermediate and final decisions upon parties to the proceedings is governed by Rule 2010 (service) in Part 385 of this chapter. Descriptions of the Commission's organization, its methods of operation, statements of policy and interpretations, procedural and substantive rules, and amendments thereto will be filed with and published in the **Federal Register.** Commission opinions together with accompanying orders, Commission orders, and intermediate decisions will be released to the press and made available to the public promptly. Copies of Commission opinions, orders in the nature of opinions, rulemakings and selected procedural orders, and intermediate decisions which have become final are published in the *Federal Energy Guidelines* and upon payment of applicable charges, may be obtained from: Commerce Clearing House, Inc., 4025 West Peterson Avenue, Chicago, Illinois 60646. Attention: Order Department.

§ 388.104 Informal advice from Commission staff.

The Commission staff provides informal advice and assistance to the general public and to prospective applicants for licenses, certificates and other Commission authorizations. Opinions expressed by the staff do not represent the official views of the Commission, but are designed to aid the public and facilitate the accomplishment of the Commission's functions. Inquiries may be directed to the chief of the appropriate office or division. An inquiry directed to the Chief Accountant that requires a written response must be accompanied by the fee prescribed by § 381.301 of this chapter. An inquiry directed to the Office of the General Counsel for an interpretation of the Natural Gas Policy Act of 1978 must be accompanied by the fee prescribed in § 381.405 of this chapter.

§ 388.105 Procedures for press, television, radio, and photographic coverage.

(a) The Commission issues news releases on major applications, decisions, opinions, orders, rulemakings, new publications, major personnel changes, and other matters of general public interest. Releases are issued by and available to the media from the Office of External Affairs. Releases may be obtained by the public through the Public Reference Room.

(b) Press, television, radio and photographic coverage of Commission proceedings is permitted as follows:

(1) Press tables are located in each hearing room, and all sessions of hearings are open to the press, subject to standards of conduct applicable to all others present;

(2) Television, movie and still cameras, and recording equipment are permit-

ted in hearing rooms prior to the opening of a hearing or oral arguments, and during recesses, upon prior arrangement with the Commission or presiding administrative law judge. All equipment must be removed from the room before hearings or oral arguments begin or resume;

(3) Television, movie and still cameras, and recording equipment may not be used while hearings and oral arguments before administrative law judges are in progress;

(4) Television and press cameras and recording equipment may be used at Commission press conferences under prior arrangement with the Office of External Affairs, provided their use does not interfere with the orderly conduct of the press conference;

(5) Regulations pertaining to the use of television, movie and still cameras, and recording equipment in connection with the Commission's open public meetings under the Government in the Sunshine Act are found in § 375.203 of this chapter.

§ 388.106 Requests for Commission records available in the Public Reference Room.

(a) A Public Reference Room is maintained at the Commission's headquarters and is open during regular business hours as provided in § 375.101(c) of this chapter. Documents may be obtained in person or in writing from the Public Reference Room by reasonably describing the records sought.

(b) The public records of the Commission that are available for inspection and copying upon request in the Public Reference Room include:

(1) Applications, declarations, complaints, petitions, and other papers seeking Commission action;

(2) Financial, statistical, and other reports to the Commission, power system statements of claimed cost of licensed projects, original cost and reclassification studies, proposed accounting entries, certificates of notification [under section 204(e) of the Federal Power Act], rates or rate schedules and related data and concurrences, and other filings and submittals to the Commission in compliance with the requirements of any statute, executive order, or Commission rule, regulation, order, license, or permit;

(3) Answers, replies, responses, objections, protests, motions, stipulations, exceptions, other pleadings, notices, certificates, proofs of service, transcripts of oral arguments, and briefs in any matter or proceeding;

(4) Exhibits, attachments and appendices to, amendments and corrections of, supplements to, or transmittals or withdrawals of any of the foregoing;

(5) All parts of the formal record in any matter or proceeding set for formal or statutory hearing, and any Commission correspondence related thereto;

(6) Presiding officer actions, correspondence, and memoranda to or from others, with the exception of internal communications within the Office of Administrative Law Judges;

(7) Commission orders, notices, findings, opinions, determinations, and other actions in a matter or proceeding;

(8) Commission correspondence relating to any furnishing of data or information, except to or by another branch, department, or agency of the Government;

(9) Commission correspondence with respect to the furnishing of data, information, comments, or recommendations to or by another branch, department, or agency of the Government where furnished to satisfy a specific requirement of a statute or where made public by that branch, department or agency;

(10) Staff reports on statements of claimed cost by licensees when such reports have been served on the licensee;

(11) Commission correspondence on interpretation of the Uniform System of Accounts and letters on such interpretation signed by the Chief Accountant and sent to persons outside the Commission;

(12) Commission correspondence on the interpretation or applicability of any statute, rule, regulation, order, license, or permit issued or administered by the Commission, and letters of opinion on that subject signed by the General Counsel and sent to persons outside the Commission;

(13) Copies of the filings, certifications, pleadings, records, briefs, orders, judgments, decrees, and mandates in court proceedings to which the Commission is a party and the correspondence with the courts or clerks of court;

(14) The Commission's Directives System;

(15) The Commission's opinions, decisions, orders and rulemakings;

(16) Reports, decisions, maps, and other information on electric power and natural gas industries;

(17) Subject index of major Commission actions;

(18) Annual report to Congress in which the Commission's operations during a past fiscal year are described; and

(19) Commission correspondence relating to the foregoing.

(c) For purposes of this section,

(1) "Commission correspondence" includes written communications and enclosures received from others outside the staff and intended for the Commission or sent to others outside the staff and signed by the Chairman, a Commissioner, the Secretary, the Executive Director, or other authorized official, except those which are personal.

(2)"Formal record" includes:

(i) Filings and submittals in a matter or proceeding,

(ii) Any notice or Commission order initiating the matter or proceeding, and

(iii) If a hearing is held, the designation of the presiding officer, transcript of hearing, exhibits received in evidence, exhibits offered but not received in evidence, offers of proof, motions, stipulations, subpoenas, proofs of service, references to the Commission, and determinations made by the Commission thereon, certifications to the Commission, and anything else upon which action of the presiding officer or the Commission may be based.

The "formal record" does not include proposed testimony or exhibits not offered or received in evidence.

(3) "Matter or proceeding" means the Commission's elucidation of the relevant facts and applicable law, consideration thereof, and action thereupon with respect to a particular subject within the Commission's jurisdiction, initiated by a filing or submittal or a Commission notice or order.

§ 388.107 Commission records exempt from public disclosure.

The following records are exempt from disclosure.

(a)(1) Records specifically authorized under criteria established by an Executive order to be kept secret in the interest of natural defense or foreign policy, and

(2) Those records [that] are in fact properly classified pursuant to such Executive order;

(b) Records related solely to the internal personnel rules and practices of an agency;

(c) Records specifically exempted from disclosure by statute, provided that such statute:

(1) Requires that the matters be withheld from the public in such a manner as to leave no discretion on the issue, or

(2) Establishes particular criteria for withholding or refers to particular types of matters to be withheld;

(d) Trade secrets and commercial or financial information obtained from a person and privileged or confidential;

(e) Interagency or intraagency memoranda or letters which would not be available by law to a party other than an agency in litigation with the agency;

(f) Personnel and medical files and similar files the disclosure of which would constitute a clearly unwarranted invasion of personal privacy;

(g) Records or information compiled for law enforcement purposes, but only to the extent that the production of such law enforcement records or information:

(1) Could reasonably be expected to interfere with enforcement proceedings,

(2) Would deprive a person of a right to a fair trial or an impartial adjudication,

(3) Could reasonably be expected to constitute an unwarranted invasion of personal privacy,

(4) Could reasonably be expected to disclose the identity of a confidential source, including a state, local, or foreign agency or authority or any private institution which furnished information on a confidential basis, and, in the case of a record or information compiled by a criminal law enforcement authority in the course of a criminal investigation, or by an agency conducting a lawful national security intelligence investigation, information furnished by a confidential source,

(5) Would disclose techniques and procedures for law enforcement investigations or prosecutions, or would disclose guidelines for law enforcement

investigations or prosecutions if such disclosure could reasonably be expected to risk circumvention of the law, or

(6) Could reasonably be expected to endanger the life or physical safety of any individual;

(h) Geological and geophysical information and data, including maps, concerning wells.

§ 388.108 Requests for Commission records not available through the Public Reference Room (FOIA requests).

(a)(1) Except as provided in paragraph (a)(2), of this section, a person may request access to Commission records that are not available through the Public Reference Room by using the following procedures:

(i) The request must be in writing, addressed to the Director of Public Affairs, and clearly marked "Freedom of Information Act Request."

(ii) The request must include:

(A) A statement by the requester of a willingness to pay a reasonable fee or fees not to exceed a specific amount, or

(B) A request for waiver or reduction of fees.

(iii) The request must identify the category of the request, consistent with the definitions provided in § 388.109(b) (1).

(2) A request that fails to provide the identification required in paragraph (a)(1)(iii) of this section will not be processed until the Director of Public Affairs can ascertain the requester's category.

(3) A request for records received by the Commission not addressed and marked as indicated will be so addressed and marked by Commission personnel as soon as it is properly identified, and forwarded immediately to the Director of Public Affairs.

(4) Requests made pursuant to this section will be considered to be received upon actual receipt and, if necessary, categorization by the Director of Public Affairs.

(b)(1) Except as provided in paragraph (b)(2) of this section, within 10 working days after receipt of the request, the Director of Public Affairs will determine whether to comply with the request for agency records and will notify the person making the request of the determination and the reasons for a decision to deny the request, and of the right of the requester to appeal any adverse determination in writing to the General Counsel or General Counsel's designee.

(2) Pursuant to § 388.110, the time limit for an initial determination may be extended by up to 10 working days.

(c) The procedure for appeal of denial of a request for Commission records is set forth in § 388.110.

§ 388.109 Fees for record requests.

(a) *Fees for records available through the Public Reference Room.*—(1) *General rule.* The fee for finding and duplicating records available in the Commission's

Public Reference Room will vary depending on the size and complexity of the request. A schedule of fees for such services is prescribed annually. A person can obtain a copy of the schedule of fees in person or by mail from the Public Reference Room. Copies of documents also may be made on self-service duplicating machines located in the Public Reference Room. In addition, copies of data extracted from the Commission's files through electronic media are available on a reimbursable basis, upon written request to the Public Reference Room.

(2) Stenographic reports of Commission hearings are made by a private contractor. Interested persons may obtain copies of public hearing transcripts from the contractor at prices set in the contract, or through the search and duplication service noted above. Copies of the contract are available for public inspection in the Public Reference Room.

(3) Copies of transcripts, electronic recordings, or minutes of Commission meetings closed to public observation containing material nonexempt pursuant to § 375.206(f) of this chapter are also available at the actual cost of duplication or transcription.

(4) The public may purchase hard copies of certain documents from the Commission's Records Information Management System (RIMS). The fee is 15 cents per page. There will be no charge for requests consisting of 10 or fewer pages.

(5) Except for requests for certification by Government agencies, certification of copies of official Commission records must be accompanied by a fee of $5.00 per document. Inquiries and orders may be made to the Public Reference Room in person or by mail.

(b) *Fees for records not available through the Public Reference Room (FOIA requests).* The cost of duplication of records not available in the Public Reference Room will depend on the number of documents requested, the time necessary to locate the documents requested, and the category of the persons requesting the records. The procedures for appeal of requests for fee waiver or reduction are set forth in § 388.110.

(1) *Definitions.* For the purpose of paragraph (b) of this section:

(i) "Commercial use" request means a request from or on behalf of one who seeks information for a use or purpose that furthers commercial, trade, or profit interests as these phrases are commonly known or have been interpreted by the courts in the context of the Freedom of Information Act;

(ii) "Educational institution" refers to a preschool, a public or private elementary or secondary school, an institution of graduate higher education, an institution of undergraduate higher education, an institution of professional education, and an institution of vocational education, which operates a program of scholarly research;

(iii) "Noncommercial scientific institution" refers to an institution that is not operated on a commercial basis and which is operated solely for the purpose of conducting scientific research the results of which are not intended to promote any particular product or industry;

(iv) "Representative of the news media" refers to any person actively gathering news for an entity that is organized and operated to publish or broadcast news to the public. The term "news" means information that is about current events or that would be of current interest to the public. Examples of news media entities include television or radio stations broadcasting to the public at large, and publishers of periodicals (but only in those instances when the periodicals can qualify as disseminations of "news") who make their products available for purchase or subscription by the general public. These examples are not intended to be all-inclusive. Moreover, as traditional methods of news delivery evolve (e.g., electronic dissemination of newspapers through telecommunications services), such alternative media may be included in this category. A "freelance" journalist may be regarded as working for a news organization if the journalist can demonstrate a solid basis for expecting publication through that organization, even though the journalist is not actually employed by the news organization. A publication contract would be the clearest proof, but the Commission may also look to the past publication record of a requester in making this determination.

(2) *Fees.* (i) If documents are requested for commercial use, the Commission will charge the employee's hourly pay rate plus 16 percent for benefits for document search time and for document review time, and 15 cents per page for duplication. Commercial use requests are not entitled to two hours of free search time or 100 free pages of reproduction of documents.

(ii) If documents are not sought for commercial use and the request is made by an educational or noncommercial scientific institution, whose purpose is scholarly or scientific research, or a representative of the news media, the Commission will charge 15 cents per page for duplication. There is no charge for the first 100 pages.

(iii) For a request not described in paragraphs (b)(2)(i) or (ii) of this section the Commission will charge the employee's hourly pay rate plus 16 percent for benefits for document search time and document review time, and 15 cents per page for duplication. There is no charge for the first 100 pages of reproduction and the first two hours of search time will be furnished without charge.

(iv) If documents are mailed, requesters will be charged postage based on the current postage rate.

(v) The Commission, or its designee, may establish minimum fees below which no charges will be collected, if it determines that the costs of routine collection and processing of the fees are likely to equal or exceed the amount of the fees. If total fees assessed by Commission staff for a Freedom of Information Act request are less than the appropriate threshold, the Commission may not charge the requesters.

(vi) Payment of fees must be by check or money order made payable to the U.S. Treasury.

(vii) Requesters may not file multiple requests at the same time, each seeking portions of a document or documents, solely in order to avoid payment of fees. When the Commission reasonably believes that a requester, or a group of requesters acting in concert, is attempting to break a request down into a series of requests for the purpose of evading assessment of fees, the Commission may aggregate any such requests and charge the requester accordingly. The Commission will not aggregate multiple requests on unrelated subjects from a requester.

(3) *Fees for unsuccessful search.* The Commission may assess charges for time spent searching, even if it fails to locate the records, or if records located are determined to be exempt from disclosure. If the Commission estimates that search charges are likely to exceed $25, it will notify the requester of the estimated amount of search fees, unless the requester has indicated in advance willingness to pay fees as high as those anticipated. The requester can meet with Commission personnel with the object of reformulating the request to meet his or her needs at a lower cost.

(4) *Interest—notice and rate.* The Commission will assess interest charges on an unpaid bill starting on the 31st day following the day on which the billing was sent. Interest will be at the rate prescribed in 31 U.S.C. 3717 and will accrue from the date of the billing.

(5) *Advance payments.* The Commission will require a requester to make an advance payment, *i.e.*, payment before work is commenced or continued on a request, if:

(i) The Commission estimates or determines that allowable charges that a requester may be required to pay are likely to exceed $250. The Commission will notify the requester of the estimated cost and either require satisfactory assurance of full payment where the requester has a history of prompt payment of fees, or require advance payment of the charges if a requester has no history of payment; or

(ii) A requester has previously failed to pay a fee charged in a timely fashion, the Commission will require the requester to pay the full amount owed plus any applicable interest, and to make an advance payment of the full amount of the estimated fee before the Commission will begin to process a new request or a pending request from that requester.

(iii) When the Commission requires advance payment under this paragraph, the administrative time limits prescribed in this part will begin only after the Commission has received the fee payments described above.

(6) *Fee reduction or waiver.* (i) Any fee described in paragraph (b) of this section may be reduced or waived if the requester demonstrates that disclosure of the information sought is:

(A) In the public interest because it is likely to contribute significantly to public understanding of the operations or activities of the government, and

(B) Not primarily in the commercial interest of the requester.

(ii) The Commission will consider the following criteria to determine the public interest standard:

(A) Whether the subject of the requested records concerns the operations or activities of the government;

(B) Whether the disclosure is likely to contribute to an understanding of government operations or activities;

(C) Whether disclosure of the requested information will contribute to public understanding; and

(D) Whether the disclosure is likely to contribute significantly to public understanding of government operations or facilities.

(iii) The Commission will consider the following criteria to determine the commercial interest of the requester:

(A) Whether the requester has a commercial interest that would be furthered by the requested disclosure; and, if so

(B) Whether the magnitude of the identified commercial interest of the requester is sufficiently large, in comparison with the public interest in disclosure, that disclosure is primarily in the commercial interest of the requester.

(iv) This request for fee reduction or waiver must accompany the initial request for records and will be decided under the same procedures used for record requests.

(7) *Debt collection.* The Commission will use the authorities mandated in the Debt Collection Act of 1982, 31 U.S.C. 3711, 3716–3719 (1982), including disclosure to consumer reporting agencies and use of collection agencies, where appropriate, to encourage payment of outstanding unpaid FOIA invoices.

(8) *Annual adjustment of fees.*—(i) *Update and publication.* The Commission, by its designee, the Executive Director, will update the fees established in this section each fiscal year. The Executive Director will publish the fees in the **Federal Register.**

(ii) *Payment of updated fees.* The fee applicable to a particular Freedom of Information Act request will be the fee in effect on the date that the request is received.

§388.110 Procedure for appeal of denial of requests for Commission records not publicly available or not available through the Public Reference Room and denial of requests for fee waiver or reduction.

(a)(1) A person whose request for records or request for fee waiver or reduction is denied in whole or part may appeal that determination to the General Counsel or General Counsel's designee within 45 days of the determination. Appeals filed pursuant to this section must be in writing, addressed to the General Counsel of the Commission, and clearly marked "Freedom of Information Act Appeal." Such an appeal received by the Commission not addressed and marked as indicated in this paragraph will be so addressed and marked by Commission personnel as soon as it is properly identified and then will be forwarded to the General Counsel. Appeals taken pursuant to this paragraph will be considered to be received upon actual receipt by the General Counsel.

(2) The General Counsel or the General Counsel's designee will make a

determination with respect to any appeal within 20 working days after the receipt of such appeal. If, on appeal, the denial of the request for records or fee reduction is in whole or in part upheld, the General Counsel or the General Counsel's designee will notify the person making such request of the provisions for judicial review of that determination.

(b) In unusual circumstances, the time limits prescribed for making the initial determination pursuant to §388.108 and for deciding an appeal pursuant to this section may be extended by up to 10 working days, by the Secretary who will send written notice to the requester setting forth the reasons for such extension and the date on which a determination or appeal is expected to be dispatched. "Unusual circumstances" means:

(1) The need to search for and collect the requested records from field facilities or other establishments that are separate from the office processing the requests;

(2) The need to search for, collect, and appropriately examine a voluminous amount of separate and distinct records which are demanded in a single request; or

(3) The need for consultation, which will be conducted with all practicable speed, with another agency having a substantial interest in the determination of the request or among two or more components of the agency having substantial subject-matter interest therein.

§ 388.111 Procedures in event of subpoena.

(a)(1) The procedures specified in this section will apply to all subpoenas directed to Commission employees that relate in any way to the employees' official duties. These procedures will also apply to subpoenas directed to former Commission employees if the subpoenas seek nonpublic materials or information acquired during Commission employment. The provisions of paragraph (c) of this section will also apply to subpoenas directed to the Commission.

(2) For purposes of this section,

(i) "Employees," except where otherwise specified, includes "special government employees" and other Commission employees; and

(ii) "Nonpublic" includes any material or information which is exempt from availability for public inspection and copying;

(iii) "Special government employees" includes consultants and other employees as defined by section 202 of Title 18 of the United States Code;

(iv) "Subpoena" means any compulsory process in a case or matter, including a case or matter to which the Commission is not a party.

(b) Any employee who is served with a subpoena must promptly advise the General Counsel of the Commission of the service of the subpoena, the nature of the documents or information sought, and all relevant facts and circumstances. Any former employee who is served with a subpoena that concerns nonpublic information shall promptly advise the General Counsel of the Commission of the service of the subpoena, the nature of the documents or information sought, and all relevant facts and circumstances.

(c) A party causing a subpoena to be issued to the Commission or any employee or former employee of the Commission must furnish a statement to the General Counsel of the Commission. This statement must set forth the party's interest in the case or matter, the relevance of the desired testimony or documents, and a discussion of whether the desired testimony or documents are reasonably available from other sources. If testimony is desired, the statement must also contain a general summary of the testimony and a discussion of whether Commission records could be produced and used in lieu of testimony. Any authorization for testimony will be limited to the scope of the demand as summarized in such statement.

(d) Commission records or information which are not part of the public record will be produced only upon authorization by the Commission.

(e) The Commission or its designee will consider and act upon subpoenas under this section with due regard for statutory restrictions, the Commission's Rules of Practice and Procedure, and the public interest, taking into account factors such as applicable privileges including the deliberative process privilege; the need to conserve the time of employees for conducting official business; the need to avoid spending the time and money of the United States for private purposes; the need to maintain impartiality between private litigants in cases where a substantial government interest is not involved; and the established legal standards for determining whether justification exists for the disclosure of confidential information and records.

(f) The Commission authorizes the General Counsel or the General Counsel's designee to make determinations under this section.

§ 388.112 Requests for privileged treatment of documents submitted to the Commission.

(a) *Scope.* Any person submitting a document to the Commission may request privileged treatment by claiming that some or all of the information contained in a particular document is exempt from the mandatory public disclosure requirements of the Freedom of Information Act, 5 U.S.C. 552, and should otherwise be withheld from public disclosure.

(b) *Procedures.* A person claiming that information is privileged under paragraph (a) of this section must file:

(1) A written statement requesting privileged treatment for some or all of the information in a document, and the justification for nondisclosure of the information;

(2) The original document, boldly indicating on the front page "Contains Privileged Information—Do Not Release" and identifying within the document the information for which the privileged treatment is sought;

(3) Fourteen copies of the document without the information for which privileged treatment is sought, and with a statement indicating that information has been removed for privileged treatment;

(4) The name, title, address, telephone number, and telecopy information of the person or persons to be contacted regarding the request for privileged treatment of documents submitted to the Commission.

(c) *Effect of privilege claim.*—(1) *For documents filed with the Commission.* (i) The Secretary of the Commission will place documents for which privileged treatment is sought in accordance with paragraph (b)(2) of this section in a nonpublic file, while the request for confidential treatment is pending. By placing documents in a nonpublic file, the Commission is not making a determination on any claim for privilege. The Commission retains the right to make determinations with regard to any claim of privilege, and the discretion to release information as necessary to carry out its jurisdictional responsibilities.

(ii) The Secretary of the Commission will place the request for privileged treatment described in paragraph (b)(1) of this section and a copy of the original document described in paragraph (b)(3) of this section in a public file, while the request for privilege treatment is pending.

(2) *For documents submitted to Commission staff.* The notification procedures of paragraphs (d) (e) and (f) of this section will be followed by staff before making a document public.

(d) *Notification of request and opportunity to comment.* When a FOIA requester seeks a document for which privilege is claimed, the Commission official who will decide whether to make the document public will notify the person who submitted the document and give the person an opportunity (at least five days) in which to comment in writing on the request. A copy of this notice will be sent to the FOIA requester.

(e) *Notification before release.* Notice of a decision by the Director of the Division of Public Affairs, the Chairman of the Commission, the General Counsel or General Counsel's designee, a presiding officer in a proceeding under Part 385 of this chapter, or any other appropriate official to deny a claim of privilege, in whole or in part, will be given to any person claiming that information is privileged no less than five days before public disclosure. The notice will briefly explain why the person's objections to disclosure are not sustained by the Commission. A copy of this notice will be sent to the FOIA requester.

(f) *Notification of suit in Federal courts.* When a FOIA requester brings suit to compel disclosure of confidential commercial information, the Commission will notify the person who submitted documents containing confidential commercial information of the suit.

Appendix G

FERC's 1989–1999 Relicensing Workload

A FERC hydroelectric license is for a limited term. When the original license expires, FERC must decide whether, to whom, and on what conditions to grant a new license for an existing hydroelectric project. Over 200 nonfederal hydroelectric projects will come up for relicensing by 1993. FERC's latest list of all the projects scheduled to be relicensed between 1989 and 1999 is set forth on the following pages.

1989–1999 Relicensing Workload
March 29, 1989

State	*Expiration*	*FERC No.*		*Proj./Develop. Name*	*Stream/River*	*River Basin*	*Applicant's Name*
AK	95/04/30	1922	A	Beaver Falls	George Inlet	Ketchikan Group	City of Ketchikan
AK	95/04/30	1922	B	Lower Silvis	George Inlet	Ketchikan Group	City of Ketchikan
AK	95/04/30	1922	C	Upper Silvis	George Inlet	Ketchikan Group	City of Ketchikan
AL	93/12/31	2407	A	Yates	Tallapoosa	Alabama-Coosa	Alabama Pwr Co
AL	93/12/31	2408	A	Thurlow	Tallapoosa	Alabama-Coosa	Alabama Pwr Co
AZ	94/12/31	2069	A	Childs	Fossil Cr/Verde Ri	Gila	Arizona Pub Serv Co
AZ	94/12/31	2069	B	Irving	Fossil Cr/Verde Ri	Gila	Arizona Pub Serv Co
CA	96/04/30	1932	A	Lytle Creek	Santa Ana	"Minor River Basin"	Southern Cal Edison
CA	96/04/30	1933	A	Santa Ana 2	Santa Ana	"Minor River Basin"	Southern Cal Edison
CA	96/04/30	1933	B	Santa Ana 1	Santa Ana	"Minor River Basin"	Southern Cal Edison
CA	96/04/30	1934	A	Mill Creek 23	Santa Ana	"Minor River Basin"	Southern Cal Edison
CA	96/04/30	1934	B	Mill Creek 3	Santa Ana	"Minor River Basin"	Southern Cal Edison
CA	91/07/31	1403	A	Narrows	Yuba	Feather	Pacific Gas & Elec
CA	93/12/31	2290	A	Kern River 3	N Fk Kern	Kern	Southern Cal Edison
CA	96/04/30	1930	A	Kern 1	Kern	Kern	Southern Cal Edison
CA	89/04/30	1333	A	Tule River	N Fk M Fk Tule	Kings Kaweah	Pacific Gas & Elec
CA	95/12/31	2687	A	Pit 1	Pit	Pit	Pacific Gas & Elec
CA	99/02/28	2017	A	Big Creek 4	San Joaquin	San Joaquin	Southern Cal Edison
CA	95/12/31	2699	A	Angels	Clovey Creek	Stanislaus	Pacific Gas & Elec
CA	96/11/05	2019	A	Murphys	Clovey Creek	Stanislaus	Pacific Gas & Elec
CA	89/04/03	1354	A	Wishon A G	San Joaquin	Willow	Pacific Gas & Elec

CA	89/04/30	1354	B	San Joaquin 1A	San Joaquin	Willow	Pacific Gas & Elec
CA	89/04/30	1354	C	San Joaquin 2	Willow	Willow	Pacific Gas & Elec
CA	89/04/30	1354	D	San Joaquin 3	N Fk Willow	Willow	Pacific Gas & Elec
CA	89/04/30	1354	E	Crane Valley	N Fk Willow	Willow	Pacific Gas & Elec
CA	89/04/30	1354	F	Browns Creek	S Fk Willow	Willow	Pacific Gas & Elec
CO	93/12/31	2187	A	Georgetown	S Clear Cr	Platte	Pub Serv Co of CO
CO	93/12/31	2275	A	Salida 2	Arkansas	Upper Arkansas	Pub Serv Co of CO
CO	93/12/31	2275	B	Salida 1	Arkansas	Upper Arkansas	Pub Serv Co of CO
CT	93/12/31	2441	A	Second Street	Shetucket	Thames	City of Norwich
CT	93/12/31	2508	A	Tenth Street	Shetucket	Thames	City of Norwich
GA	93/12/31	2336	A	Lloyd Shoals	Ocmulgee	Altamaha	Georgia Pwr Co
GA	97/08/31	1951	A	Sinclair	Oconee	Altamaha	Georgia Pwr Co
GA	93/12/31	2341	A	Langdale	Chattahoochee	Apalachicola	Georgia Pwr Co
GA	93/12/31	2350	A	Riverview 5	Chattahoochee	Apalachicola	Georgia Pwr Co
GA	93/12/31	2354	A	Yonah	Tugaloo	Savannah	Georgia Pwr Co
GA	93/12/31	2354	B	Tugaloo	Tugaloo	Savannah	Georgia Pwr Co
GA	93/12/31	2354	C	Tallulah Falls	Tallulah	Savannah	Georgia Pwr Co
GA	93/12/31	2354	D	Mathis & Terror	Tallulah	Savannah	Georgia Pwr Co
GA	93/12/31	2354	E	Nacoochee	Tallulah	Savannah	Georgia Pwr Co
GA	93/12/31	2354	F	Burton	Tallulah	Savannah	Georgia Pwr Co
GA	93/12/31	2535	A	Stevens Creek	Savannah	Savannah	So Carolina E & G
ID	98/03/31	1991	A	Moyie 2	Moyie	Kootenai	City of Bonners Ferry
ID	98/03/31	1991	B	Moyie 1	Moyie	Kootenai	City of Bonners Ferry
ID	98/03/31	1991	C	Moyie 3	Moyie	Kootenai	City of Bonners Ferry
ID	97/12/23	2061	A	Lower Salmon	Snake	Upper Snake	Idaho Power Co
ID	98/02/28	1975	A	Bliss	Snake	Upper Snake	Idaho Power Co
ID	99/05/31	2777	A	Upper Salmon B	Snake	Upper Snake	Idaho Power Co

1989–1999 Relicensing Workload
March 29, 1989

State	*Expiration*	*FERC No.*		*Proj./Develop. Name*	*Stream/River*	*River Basin*	*Applicant's Name*
ID	99/05/31	2777	B	Upper Salmon B	Snake	Upper Snake	Idaho Power Co
ID	99/05/31	2778	A	Shoshone Falls	Snake	Upper Snake	Idaho Power Co
IL	93/12/31	2373	A	Rockton	Rock	Rock	So Beloit Wg & El
IL	93/12/31	2446	A	Dixon	Rock	Rock	Commonwealth Ed Co
IN	93/12/31	2579	A	Twin Branch	St Joseph	St Joseph	IN & MI Elec Co
MA	91/02/28	2386	A	Holyoke 1	Holyoke Cnl	Connecticut	City of Holyoke
MA	91/02/28	2497	A	Mt Tom Hill	Holyoke Cnl	Connecticut	Linweave, Inc
MA	91/02/28	2622	A	Turners Falls	Conn Cnl	Connecticut	Hammermill Paper Co
MA	91/02/28	2758	A	Crocker Mill	Holyoke Cnl	Connecticut	Linweave, Inc
MA	91/02/28	2766	A	Alvin Mill D	Holyoke Cnl	Connecticut	Linweave, Inc
MA	91/02/28	2768	A	Alvin Mill A	Holyoke Cnl	Connecticut	Linweave, Inc
MA	91/02/28	2770	A	Crocker Mill C	Holyoke Cnl	Connecticut	Linweave, Inc
MA	91/02/28	2771	A	Nonotuck	Holyoke Cnl	Connecticut	Linweave, Inc
MA	91/02/28	2772	A	Linweave A	Holyoke Cnl	Connecticut	Holyoke Wtr & Pwr
MA	91/02/28	2775	A	Linweave D	Holyoke Cnl	Connecticut	Linweave, Inc
MA	93/12/31	2323	A	Deerfield 2	Deerfield	Connecticut	New England Pwr Co
MA	93/12/31	2323	B	Deerfield 3	Deerfield	Connecticut	New England Pwr Co
MA	93/12/31	2323	C	Deerfield 4	Deerfield	Connecticut	New England Pwr Co
MA	93/12/31	2323	D	Deerfield 5	Deerfield	Connecticut	New England Pwr Co
MA	93/12/31	2323	E	Sherman	Deerfield	Connecticut	New England Pwr Co
MA	93/12/31	2334	A	Gardner Falls	Deerfield	Connecticut	Western MA Elect
MA	93/12/31	2608	A	West Springfield	Westfield	Connecticut	James R Paper Co
MA	99/08/31	2004	A	Hadley Falls	Connecticut	Connecticut	Holyoke Wtr & Pwr
MA	99/08/31	2004	B	Skinner	Holyoke Cnl	Connecticut	Holyoke Wtr & Pwr
MA	99/08/31	2004	C	Boatlock Station	Holyoke Cnl	Connecticut	Holyoke Wtr & Pwr
MA	99/08/31	2004	D	Riverside	Holyoke Cnl	Connecticut	Holyoke Wtr & Pwr

MA	99/08/31	2004	E	Beebee Holbrook	Holyoke Cnl	Connecticut	Holyoke Wtr & Pwr
MA	99/08/31	2004	F	Chemical	Holyoke Cnl	Connecticut	Holyoke Wtr & Pwr
MA	99/11/30	2928	A	Merrimack	S Merrimack Cnl	Merrimack	Merrimack Paper Co
MA	99/09/30	2927	A	Aquamac	S Merrimack Cnl	Merrimack	Aquamac Corp
MD	93/12/31	2370	A	Deep Creek	Deep Creek	Monongahela	Penn Elec Co
ME	93/12/31	2283	A	Deer Rips	Androscoggin	Androscoggin	Central ME Power
ME	93/12/31	2283	B	Androscoggin No. 3	Androscoggin	Androscoggin	Central ME Power
ME	93/12/31	2283	C	Gulf Island	Androscoggin	Androscoggin	Central ME Power
ME	93/12/31	2333	A	Rumford Falls	Androscoggin	Androscoggin	Rumford Falls Pwr Co
ME	93/12/31	2333	B	Rumford Falls-Upper	Androscoggin	Androscoggin	Rumford Falls Pwr Co
ME	99/09/30	2375	A	Livermore Mill	Androscoggin	Androscoggin	Internat Paper Co
ME	99/09/30	2375	B	Jay	Androscoggin	Androscoggin	Internat Paper Co
ME	99/09/30	2375	C	Riley Mill	Androscoggin	Androscoggin	Internat Paper Co
ME	99/09/30	8277	A	Otis Mill	Androscoggin	Androscoggin	Otis Hydroelec Co
ME	93/12/31	2325	A	Weston	Kennebec	Kennebec	Central ME Power
ME	93/12/31	2329	A	Wyman	Kennebec	Kennebec	Central ME Power
ME	93/12/31	2389	A	Edwards Division	Kennebec	Kennebec	Edwards Manufct Co
ME	93/12/31	2552	A	Fort Halifax	Sebasticook	Kennebec	Central ME Power
ME	93/12/31	2555	A	Automatic (Mesal 4)	Messalonskee	Kennebec	Central ME Power
ME	93/12/31	2556	A	Union Gas (Mesal 5)	Messalonskee	Kennebec	Central ME Power
ME	93/12/31	2557	A	Rice Rips (Mesal 3)	Messalonskee	Kennebec	Central ME Power
ME	93/12/31	2559	A	Oakland (Mesal 2)	Messalonskee	Kennebec	Central ME Power
ME	93/12/31	2612	A	Flagstaff	Dead	Kennebec	Central ME Power
ME	93/12/31	2613	A	Moxie	Lake Moxie	Kennebec	Milstar Manuf Corp
ME	93/12/31	2615	A	Brassua	Moose	Kennebec	Central ME Power
ME	93/12/31	2671	A	Moosehead Lake	Kennebec	Kennebec	Kennebec Wtr Pwr Co
ME	93/12/31	2458	A	Millinocket	W Br Penobscot	Penobscot	Gr Northern Nekoosa
ME	93/12/31	2458	B	Dolby	W Br Penobscot	Penobscot	Gr Northern Nekoosa

1989–1999 Relicensing Workload
March 29, 1989

State	*Expiration*	*FERC No.*		*Proj./Develop. Name*	*Stream/River*	*River Basin*	*Applicant's Name*
ME	93/12/31	2458	C	East Millinocket	W Br Penobscot	Penobscot	Gr Northern Nekoosa
ME	93/12/31	2458	D	North Twin	W Br Penobscot	Penobscot	Gr Northern Nekoosa
ME	93/12/31	2458	E	Millinocket Dam	Millinocket	Penobscot	Gr Northern Nekoosa
ME	93/12/31	2534	A	Milford	Penobscot	Penobscot	Bangor Hydro Elec Co
ME	93/12/31	2572	A	Ripogenus	W Br Penobscot	Penobscot	Gr Northern Nekoosa
ME	93/12/31	2666	A	Medway	W Br Penobscot	Penobscot	Bangor Hydro Elec Co
ME	93/12/31	2712	A	Stillwater	Stillwater	Penobscot	Bangor Hydro Elec Co
ME	93/12/31	2519	A	North Gorham	Presumpscot	Presumpscot	Central ME Power
ME	99/09/30	2897	A	Saccarappa (Wst Br2)	Presumpscot	Presumpscot	S D Warren Co
ME	93/12/31	2527	A	Skelton	Saco	Saco	Central ME Power
ME	93/12/31	2529	A	Bonny Eagle	Saco	Saco	Central ME Power
ME	90/12/31	2368	A	Squa Pan	Squa Pan	St John	ME Pub Serv Co
ME	92/06/30	2366	A	Millinocket Lake	Millinocket	St John	ME Pub Serv Co
ME	93/12/31	2367	A	Caribou	Aroostook	St John	ME Pub Serv Co
MI	89/12/31	2610	A	Saxon Falls	Montreal	"Minor River Basin"	Northern States Pwr
MI	93/12/31	2404	A	Ninth Street	Thunder Bay	"Minor River Basin"	Alpena Power Co
MI	93/12/31	2404	B	Four Mile Dam	Thunder Bay	"Minor River Basin"	Alpena Power Co
MI	93/12/31	2404	C	Norway Point	Thunder Bay	"Minor River Basin"	Alpena Power Co
MI	93/12/31	2404	D	Hubbard Lake	Lr S Br Thunder Bay	"Minor River Basin"	Alpena Power Co
MI	93/12/31	2404	E	Upper South Dam	Up S Br Thunder Bay	"Minor River Basin"	Alpena Power Co
MI	93/12/31	2419	A	Hillman	Thunder Bay	"Minor River Basin"	Alpena Power Co
MI	93/12/31	2451	A	Rogers	Muskegon	"Minor River Basin"	Consumers Power Co
MI	93/12/31	2452	A	Hardy	Muskegon	"Minor River Basin"	Consumers Power Co
MI	93/12/31	2468	A	Croton	Muskegon	"Minor River Basin"	Consumers Power Co
MI	93/12/31	2506	A	Escanaba 1	Escanaba	"Minor River Basin"	Mead Corp

MI	93/12/31	2506	B	Escanaba 3	Escanaba	"Minor River Basin"	Mead Corp
MI	93/12/31	2506	C	Escanaba 4	Escanaba	"Minor River Basin"	Mead Corp
MI	93/12/31	2587	A	Superior Falls	Montreal	"Minor River Basin"	Northern States Pwr
MI	93/12/31	2436	A	Foote	Au Sable	Au Sable	Consumers Power
MI	93/12/31	2447	A	Alcona	Au Sable	Au Sable	Consumers Power
MI	93/12/31	2448	A	Mio	Au Sable	Au Sable	Consumers Power
MI	93/12/31	2449	A	Loud	Au Sable	Au Sable	Consumers Power
MI	93/12/31	2450	A	Cooke	Au Sable	Au Sable	Consumers Power
MI	93/12/31	2453	A	Five Channels	Au Sable	Au Sable	Consumers Power
MI	93/12/31	2580	A	Tippy	Manistee	Manistee	Consumers Power Co
MI	93/12/31	2599	A	Hodenpyl	Manistee	Manistee	Consumers Power Co
MI	93/06/30	2394	A	Chalk Hill	Menominee	Menominee	WI Elec Pwr Co
MI	93/12/31	2357	A	White Rapids	Menominee	Menominee	WI Elec Pwr Co
MI	93/12/31	2431	A	Brule Island	Brule	Menominee	WI Elec Pwr Co
MI	93/12/31	2433	A	Grand Rapids	Menominee	Menominee	WI Pub Serv Corp
MI	93/12/31	2471	A	Sturgeon	Sturgeon	Menominee	WI Elec Pwr Co
MI	98/02/28	1980	A	Quinnesec Falls	Menominee	Menominee	WI Elec Pwr Co
MI	98/02/28	1980	B	Big Quinnesec Falls	Menominee	Menominee	WI Elec Pwr Co
MI	93/12/31	2551	A	Buchanan	St Joseph	St Joseph	IN & MI Elec Co
MI	93/12/31	2402	A	Prickett	Sturgeon	Sturgeon	Up Peninsula Pwr Co
MN	93/12/31	2454	A	Sylvan	Crow Wing	Crow Wing	MN Pwr & Lt Co
MN	97/05/11	2663	A	Pillager	Crow Wing	Crow Wing	MN Pwr & Lt Co
MN	93/12/31	2361	A	Prairie River	Prairie	Mississippi–M Stem	MN Pwr & Lt Co
MN	93/12/31	2362	A	Grand Rapids	Mississippi	Mississippi–M Stem	Blandin Paper Co
MN	93/12/31	2532	B	Little Falls 2	Mississippi	Mississippi–M Stem	MN Pwr & Lt Co
MN	93/12/31	2533	A	Brainerd	Mississippi	Mississippi–M Stem	Potlatch Corp
MN	93/12/31	2532	A	Little Falls 1	Mississippi	Mississippi–M Stem	MN Pwr & Lt Co

1989–1999 Relicensing Workload
March 29, 1989

State	Expiration	FERC No.		Proj./Develop. Name	Stream/River	River Basin	Applicant's Name
MN	93/12/31	2360	A	Fond Du Lac	St Louis	St Louis	MN Pwr & Lt Co
MN	93/12/31	2360	B	Thompson	St Louis	St Louis	MN Pwr & Lt Co
MN	93/12/31	2360	C	Scanlon	St Louis	St Louis	MN Pwr & Lt Co
MN	93/12/31	2360	D	Knife Falls	St Louis	St Louis	MN Pwr & Lt Co
MN	93/12/31	2360	E	Fish Lake	Beaver	St Louis	MN Pwr & Lt Co
MN	93/12/31	2360	F	Rice Lake	Beaver	St Louis	MN Pwr & Lt Co
MN	93/12/31	2360	G	Island Lake	Cloquet	St Louis	MN Pwr & Lt Co
MN	93/12/31	2360	H	Boulder Lake	Otter	St Louis	MN Pwr & Lt Co
MN	93/12/31	2360	I	Whiteface	Whiteface	St Louis	MN Pwr & Lt Co
MN	93/12/31	2363	A	Cloquet	St Louis	St Louis	Potlatch Corp
MO	93/01/31	2561	A	Niangua	Niangua	Osage	Sho Me Pwr Co
MO	93/08/31	2221	A	Ozark Beach	White	Upper White	Empire Dist Elec Co
MT	93/12/31	2543	A	Milltown	Pend Oreille	Clark Fork	Montana Power Co
MT	98/11/30	2188	H	Madison 2	Madison	Madison	Montana Power Co
MT	98/11/30	2188	I	Hebgen	Madison	Madison	Montana Power Co
MT	98/11/30	2188	A	Morony	Missouri	Missouri–Main Stem	Montana Power Co
MT	98/11/30	2188	B	Ryan	Missouri	Missouri–Main Stem	Montana Power Co
MT	98/11/30	2188	C	Cochrane	Missouri	Missouri–Main Stem	Montana Power Co
MT	98/11/30	2188	D	Rainbow	Missouri	Missouri–Main Stem	Montana Power Co
MT	98/11/30	2188	E	Black Eagle	Missouri	Missouri–Main Stem	Montana Power Co
MT	98/11/30	2188	F	Holter	Missouri	Missouri–Main Stem	Montana Power Co
MT	98/11/30	2188	G	Hauser Lake	Missouri	Missouri–Main Stem	Montana Power Co
NC	93/12/31	2607	A	Spencer Mountain	S Fk Catawba	Santee	Duke Pwr Co
NC	93/12/31	2541	A	Cascade (Brevard)	Little	Tennessee	Cascade Power Co

NH	93/12/31	2287	A	J Brodie Smith	Androscoggin	Androscoggin	Pub Serv Co NH
NH	93/12/31	2288	A	Gorham	Androscoggin	Androscoggin	Pub Serv Co NH
NH	93/12/31	2300	A	Shelburne	Androscoggin	Androscoggin	Brown–NH, Inc
NH	93/12/31	2311	A	Gorham	Androscoggin	Androscoggin	James R–NH Elec, Inc
NH	93/12/31	2326	A	Cross	Androscoggin	Androscoggin	James R–NH Elec, Inc
NH	93/12/31	2327	A	Cascade	Androscoggin	Androscoggin	James R–NH Elec, Inc
NH	93/12/31	2422	A	Sawmill	Androscoggin	Androscoggin	James R–NH Elec, Inc
NH	93/12/31	2423	A	Riverside	Androscoggin	Androscoggin	James R–NH Elec, Inc
NH	90/12/31	2392	A	Gilman	Connecticut	Connecticut	Georgia–Pacific
NH	93/12/31	2456	A	Ayers Island	Pemigewasset	Merrimack	Pub Serv Co NH
NV	91/09/30	1746	A	Leidy Cr (Circle L)	Leidy	Carson	Four Rent, Inc
NY	93/12/31	2442	A	Watertown	Black	Black	City of Watertown
NY	93/12/31	2538	A	Beebee Island	Black	Black	Niagara Mohawk Co
NY	93/12/31	2569	A	Sewalls Island	Black	Black	Niagara Mohawk Co
NY	93/12/31	2569	B	Black River	Black	Black	Niagara Mohawk Co
NY	93/12/31	2569	C	Kamargo	Black	Black	Niagara Mohawk Co
NY	93/12/31	2569	D	Deferiet	Black	Black	Niagara Mohawk Co
NY	93/12/31	2569	E	Herrings	Black	Black	Niagara Mohawk Co
NY	93/12/31	2645	A	High Falls	Beaver	Black	Niagara Mohawk Co
NY	93/12/31	2645	B	Belfort	Beaver	Black	Niagara Mohawk Co
NY	93/12/31	2645	C	Taylorville	Beaver	Black	Niagara Mohawk Co
NY	93/12/31	2645	D	Elmer	Beaver	Black	Niagara Mohawk Co
NY	93/12/31	2645	E	Effley	Beaver	Black	Niagara Mohawk Co
NY	93/12/31	2645	F	Soft Maple	Beaver	Black	Niagara Mohawk Co
NY	93/12/31	2645	G	Eagle	Beaver	Black	Niagara Mohawk Co
NY	93/12/31	2645	H	Moshier	Beaver	Black	Niagara Mohawk Co
NY	91/06/30	2424	A	Hydraulic Race	Barge Cnl	Genesee	Niagara Mohawk Co
NY	93/12/31	2582	A	Station 2	Genesee	Genesee	Rochester Gas & Elec

1989–1999 Relicensing Workload
March 29, 1989

State	*Expiration*	*FERC No.*		*Proj./Develop. Name*	*Stream/River*	*River Basin*	*Applicant's Name*
NY	93/12/31	2582	B	Station 2	Genesee	Genesee	Rochester Gas & Elec
NY	93/12/31	2583	A	Station 5	Genesee	Genesee	Rochester Gas & Elec
NY	93/12/31	2584	A	Station 26	Genesee	Genesee	Rochester Gas & Elec
NY	93/12/31	2596	A	Station 160	Genesee	Genesee	Rochester Gas & Elec
NY	93/12/31	2318	A	EJ West	Sacandaga	Hudson	Niagara Mohawk Co
NY	93/12/31	2385	A	Glens Falls	Hudson	Hudson	Finch Pruyn & Co
NY	93/12/31	2482	D	Sherman Island	Hudson	Hudson	Niagara Mohawk Co
NY	93/12/31	2482	E	Spier Falls	Hudson	Hudson	Niagara Mohawk Co
NY	93/12/31	2487	A	Hoosic Falls	Hoosic	Hudson	Hydro-Power, Inc
NY	93/12/31	2539	A	School St	Mohawk	Hudson	Niagara Mohawk Co
NY	93/12/31	2554	A	Feeder Dam	Hudson	Hudson	Moreau Manufct Corp
NY	93/12/31	2616	A	Schaghticoke	Hoosic	Hudson	Niagara Mohawk Co
NY	93/12/31	2616	B	Johnsonville	Hoosic	Hudson	Niagara Mohawk Co
NY	93/12/31	6032	A	Mechanicville	Hudson	Hudson	Niagara Mohawk Co
NY	93/12/31	2438	A	Seneca Falls (Lk 10)	Seneca Cnl	Oswego	NY St Elec & Gas Co
NY	93/12/31	2438	B	Waterloo (Lock 4)	Seneca Cnl	Oswego	NY St Elec & Gas Co
NY	93/12/31	2474	A	Varick	Oswego	Oswego	Niagara Mohawk Co
NY	93/12/31	2474	B	Minetto	Oswego	Oswego	Niagara Mohawk Co
NY	93/12/31	2474	C	Fulton	Oswego	Oswego	Niagara Mohawk Co
NY	93/12/31	2320	A	Sugar Island	Raquette	Raquette	Niagara Mohawk Co
NY	93/12/31	2320	B	Hannawa	Raquette	Raquette	Niagara Mohawk Co
NY	93/12/31	2320	C	Colton	Raquette	Raquette	Niagara Mohawk Co
NY	93/12/31	2320	D	Higley	Raquette	Raquette	Niagara Mohawk Co
NY	93/12/31	2330	A	Raymondville	Raquette	Raquette	Niagara Mohawk Co
NY	93/12/31	2330	B	Norfolk	Raquette	Raquette	Niagara Mohawk Co
NY	93/12/31	2330	C	East Norfolk	Raquette	Raquette	Niagara Mohawk Co

NY	93/12/31	2330	D	Norwood	Raquette	Raquette	Potsdam Paper Corp
OR	93/12/31	2643	A	Bend	Deschutes	Deschutes	PC/UP&L Merging Corp
OR	96/06/29	1986	A	Rock Creek	Powder River	Lower Snake	CP National Corp
OR	97/01/29	1927	A	Soda Springs	N Umpqua	Umpqua	PC/UP&L Merging Corp
OR	97/01/29	1927	B	Slice Creek	N Umpqua	Umpqua	PC/UP&L Merging Corp
OR	97/01/29	1927	C	Fish Creek	N Umpqua	Umpqua	PC/UP&L Merging Corp
OR	97/01/29	1927	D	Toketee	N Umpqua	Umpqua	PC/UP&L Merging Corp
OR	97/01/29	1927	E	Clearwater 2	Clearwater	Umpqua	PC/UP&L Merging Corp
OR	97/01/29	1927	F	Clearwater 1	Clearwater	Umpqua	PC/UP&L Merging Corp
OR	97/01/29	1927	G	Lemolo 2	N Umpqua	Umpqua	PC/UP&L Merging Corp
OR	97/01/29	1927	H	Lemolo 1	N Umpqua	Umpqua	PC/UP&L Merging Corp
OR	93/12/31	2496	A	Leaburg	Mckenzie	Willamette	City of Eugene
OR	93/12/31	2510	A	Walterville	Mckenzie	Willamette	City of Eugene
SC	93/12/31	2315	A	Neal Shoals 5	Broad	Santee	So Carolina E & G
SC	93/12/31	2331	A	99 Islands	Broad	Santee	Duke Pwr Co
SC	93/12/31	2332	A	Gaston Shoals	Broad	Santee	Duke Pwr Co
SC	93/12/31	2406	A	Saluda	Saluda	Santee	Duke Pwr Co
SC	93/12/31	2465	A	Holidays Bridge	Saluda	Santee	Duke Pwr Co
TX	96/06/30	1952	A	Eagle Pass	Maverick Cnl	Lower Rio Grande	Maverick Co Wtr Dist
UT	91/06/22	1715	A	Spring Creek	Spring Cr/Hobble Cr	Great Salt Lake	City of Springville
UT	93/12/31	2420	A	Cutler	Bear	Great Salt Lake	PC/UP&L Merging Corp
UT	98/06/30	1994	A	Snake Creek	Snake Creek	Great Salt Lake	Heber Lt & Pwr Co
UT	93/03/31	1773	A	Yellowstone	Yellowstone	Green	Moon Lake Elec Assn
UT	93/07/30	1858	A	Beaver Upper (No 1)	Beaver	Little Salt Lake	City of Beaver
UT	90/06/30	1517	A	Monroe Upper	Monroe Creek	Sevier Lake	Monroe City Corp
VA	93/12/31	2376	A	Reusens	James	James	Appalachian Pwr Co

1989–1999 Relicensing Workload
March 29, 1989

State	*Expiration*	*FERC No.*		*Proj./Develop. Name*	*Stream/River*	*River Basin*	*Applicant's Name*
VA	93/12/31	2514	A	Buck	New	Kanawha	Appalachian Pwr Co
VA	93/12/31	2514	B	Byllesby 2	New	Kanawha	Appalachian Pwr Co
VA	93/12/31	2391	A	Warren	Shenandoah	Potomac	Potomac Edison Co
VA	93/12/31	2425	A	Luray	S Fk Shenandoah	Potomac	Potomac Edison Co
VA	93/12/31	2425	B	Newport	S Fk Shenandoah	Potomac	Potomac Edison Co
VA	93/12/31	2509	A	Shenandoah	S Fk Shenandoah	Potomac	Potomac Edison Co
VA	93/12/31	2411	A	Schoolfield	Dan	Roanoke	Sts Hydropower Ltd
VA	93/12/31	2466	A	Niagara	Roanoke	Roanoke	Appalachian Power Co
VT	93/12/31	2323	F	Harriman	Deerfield	Connecticut	New England Pwr Co
VT	93/12/31	2323	G	Searsburg	Deerfield	Connecticut	New England Pwr Co
VT	93/12/31	2323	H	Somerset	E Br Deerfield	Connecticut	New England Pwr Co
VT	93/12/31	2396	A	Pierce Mills	Passumpsic	Connecticut	Central Vt Pub Serv
VT	93/12/31	2397	A	Gage	Passumpsic	Connecticut	Central Vt Pub Serv
VT	93/12/31	2399	A	Arnold Falls	Passumpsic	Connecticut	Central Vt Pub Serv
VT	93/12/31	2400	A	Passumpsic	Passumpsic	Connecticut	Central Vt Pub Serv
VT	93/12/31	2489	A	Cavendish	Black	Connecticut	Central Vt Pub Serv
VT	93/12/31	2490	A	Taftsville	Ottauquechee	Connecticut	Central Vt Pub Serv
VT	93/12/31	2445	A	Center Rutland	Otter	Lake Champlain	Vermont Marble Co
VT	93/12/31	2513	A	Essex 19	Winooski	Lake Champlain	Green Mt Pwr Corp
VT	99/05/31	2674	A	Vergennes 9 & 9B	Otter	Lake Champlain	Green Mt Pwr Corp
VT	93/12/31	2306	A	Newport 2	Clyde	Lake Memphremagog	Citizens Utilities
VT	93/12/31	2306	B	Newport 1	Clyde	Lake Memphremagog	Citizens Utilities
VT	93/12/31	2306	C	W Charleston	Clyde	Lake Memphremagog	Citizens Utilities
VT	93/12/31	2306	D	Echo Dam	Echo Pond Outlet, Cl	Lake Memphremagog	Citizens Utilities
VT	93/12/31	2306	E	Seymour Dam	Seymour Lake, Clyde	Lake Memphremagog	Citizens Utilities

WA	93/12/31	2544	A	Meyers Falls	Colville	Kettle	Washington Wtr Pwr
WA	93/12/31	1862	A	La Grande	Nisqually	Nisqually	City of Tacoma
WA	93/12/31	1862	B	Alder	Nisqually	Nisqually	City of Tacoma
WA	94/12/31	2705	A	Newhalem Creek	Newhalem	Skagit	City of Seattle
WA	93/12/31	2493	A	Snoqualmie Falls 2	Snoqualmie	Snohomish	Puget Sound Pwr & Lt
WA	93/12/31	2493	B	Snoqualmie Falls 1	Snoqualmie	Snohomish	Puget Sound Pwr & Lt
WA	93/12/31	2342	A	Condit	White Salmon	White Salmon	PC/UP&L Merging Corp
WI	93/12/31	2444	A	White	White	"Minor River Basin"	Northern States Pwr
WI	93/12/31	2522	A	Johnson Falls	Peshtigo	"Minor River Basin"	WI Pub Serv Corp
WI	93/12/31	2523	A	Oconto Falls (Upper)	Oconto	"Minor River Basin"	Wisconsin Elec Pwr
WI	93/12/31	2525	A	Caldron Falls	Peshtigo	"Minor River Basin"	WI Pub Serv Corp
WI	93/12/31	2546	A	Sandstone Rapids	Peshtigo	"Minor River Basin"	WI Pub Serv Corp
WI	93/12/31	2560	A	Potato Rapids	Peshtigo	"Minor River Basin"	WI Pub Serv Corp
WI	93/12/31	2564	A	Orienta Falls	Iron	"Minor River Basin"	Northern States Pwr
WI	93/12/31	2581	A	Peshtigo	Peshtigo	"Minor River Basin"	WI Pub Serv Corp
WI	93/12/31	2595	A	High Falls	Peshtigo	"Minor River Basin"	WI Pub Serv Corp
WI	93/12/31	2689	A	Oconto Falls	Oconto	"Minor River Basin"	Scott Paper Co
WI	93/12/31	2390	A	Big Falls	Flambeau	Chippewa	Northern States Pwr
WI	93/12/31	2395	A	Pixley	N Fk Flambeau	Chippewa	Flambeau Paper Corp
WI	93/12/31	2421	A	Lower Hydro	N Fk Flambeau	Chippewa	Flambeau Paper Corp
WI	93/12/31	2440	A	Chippewa Falls	Chippewa	Chippewa	Northern States Pwr
WI	93/12/31	2473	A	Crowley Rapids	N Fk Flambeau	Chippewa	Flambeau Paper Corp
WI	93/12/31	2475	A	Thornapple	Flambeau	Chippewa	Northern States Pwr
WI	93/12/31	2640	A	Upper Hydro	N Fk Flambeau	Chippewa	Flambeau Paper Corp
WI	98/06/30	1982	A	Holcombe	Chippewa	Chippewa	Northern States Pwr
WI	93/12/31	2550	A	Weyauwega	Waupaca	Fox	Wisconsin Elec Pwr
WI	93/06/30	2536	A	Ltl Guinnesec Falls	Menominee	Menominee	Niagara of WI Paper

1989–1999 Relicensing Workload
March 29, 1989

State	*Expiration*	*FERC No.*		*Proj./Develop. Name*	*Stream/River*	*River Basin*	*Applicant's Name*
WI	93/12/31	2486	A	Pine	Pine	Menominee	WI Elec Pwr Co
WI	93/12/31	2347	A	Central (Janesville)	Rock	Rock	WI Pwr and Lt Co
WI	93/12/31	2348	A	Blackhawk	Rock	Rock	WI Power and Lt Co
WI	93/03/31	2711	A	Trego	Namekagon	St Croix	Northern States Pwr
WI	93/12/31	2417	A	Hayward	Namekagon	St Croix	Northern States Pwr
WI	90/06/30	1957	A	Otter Rapids	Wisconsin	Wisconsin	WI Publ Serv Corp
WI	90/06/30	1967	A	Whiting Plover	Wisconsin	Wisconsin	Kimberly Clark Corp
WI	91/06/30	1953	A	Du Bay	Wisconsin	Wisconsin	Consol Water Pwr Co
WI	93/06/30	2590	A	Wisc River Div	Wisconsin	Wisconsin	Consol Water Pwr Co
WI	93/07/31	2113	A	Wisconsin Headwaters	Wisconsin	Wisconsin	WI Valley Improv Co
WI	93/07/31	2212	A	Rothschild	Wisconsin	Wisconsin	Weyerhaeuser
WI	93/07/31	2239	A	Kings	Wisconsin	Wisconsin	Tohahawk Pwr & Pulp
WI	93/07/31	2255	A	Centralia	Wisconsin	Wisconsin	Nekoosa Papers, Inc
WI	93/07/31	2256	A	Wis Rpds 1	Wisconsin	Wisconsin	Consol Water Pwr Co
WI	93/07/31	2291	A	Port Edwards	Wisconsin	Wisconsin	Nekoosa Papers, Inc
WI	93/07/31	2292	A	Nekoosa	Wisconsin	Wisconsin	Nekoosa Papers, Inc
WI	93/12/31	2476	A	Jersey	Tomahawk	Wisconsin	WI Pub Serv Corp
WI	95/06/30	1999	A	Wausau	Wisconsin	Wisconsin	WI Pub Serv Corp
WI	98/01/31	1984	A	Castle Rock	Wisconsin	Wisconsin	WI River Pwr Co
WI	98/01/31	1984	B	Petenwell	Wisconsin	Wisconsin	WI River Pwr Co
WV	93/12/31	2459	A	Lake Lynn	Cheat	Monongahela	West Penn Power Co
WV	93/12/31	2515	A	Harpers Ferry	Potomac	Potomac	Potomac Edison Co
WY	92/11/30	1651	A	Swift Lower	Salt	Upper Snake	Swift Cr Pwr Co, Inc
WY	92/11/30	1651	B	Swift Upper	Salt	Upper Snake	Swift Cr Pwr Co, Inc
WY	99/09/30	2032	A	Strawberry	South Alt	Upper Snake	Lwr Val Pwr & Lt Co

Appendix H

How to Find and Use the Legal Documents on Which FERC Relies

In making decisions on project applications, FERC relies on legal authorities, including statutes, federal courts' opinions, and its own records in disposing of past applications. As you can see in each chapter and the accompanying footnotes, this book cites those authorities whenever they may be useful to you in understanding how FERC operates and in preparing submittals to the Commission.

There is an old saying that a lawyer doesn't know more law than a layperson: the lawyer just knows where to find relevant documents. Don't be intimidated by the citations in this book. You can locate and use the following legal materials, once you know what the citations mean.

The *United States Code* (U.S.C.) is a consolidation and codification of all general and permanent laws of the United States, organized by subject matter. The Federal Power Act is at 16 U.S.C. §§ 791a–828c, which means that it is at title 16, sections 791a through 828c.

Statutes at Large contains federal laws and congressional resolutions, organized by session of Congress but not by subject matter. A footnote citation to the Electric Consumers Protection Act of 1986 (ECPA), Pub. L. No. 99-495, 100 Stat. 1244, means that the statute is in volume 100 of *Statutes at Large,* at page 1244, and that it is the 495th law of the 99th Congress. ECPA, which mostly amended the older Federal Power Act, also is included in the *United States Code.*

The *Code of Federal Regulations* (C.F.R.) is a codification of the general and permanent rules adopted by federal departments and agencies and published in the *Federal Register.* These rules implement the statutes. The C.F.R. is organized by subject matter. FERC's rules for hydropower are at 18 C.F.R. §§ 1–141, 375–389 (general and procedural rules applicable to FERC's control of hydropower and natural gas); this citation refers to title 18 of the C.F.R., sections 1 through 141 and 375 through 389.

The *Federal Register* is the daily compilation of notices, orders, and rules of federal agencies. Notices of the filing of preliminary permit and license applications, FERC's opinions and orders on those applications, and its rules implementing its statutory directives appear in the *Federal Register.* The citation 52 Fed. Reg. 39905 refers to volume 52 of the *Federal Register* at page 39,905.

Federal Energy Regulatory Commission Reports (*FERC Reports*) contain FERC's final opinions and orders on all matters within its jurisdiction, including hydropower applications. It succeeds *Federal Power Commission Reports* (*FPC Reports*), which contain the agency's opinions and orders from 1920 until October 1, 1978, when its name was changed to FERC. *FERC Reports* are organized chronologically, mingling decisions on hydropower and natural gas matters. The citation 42 FERC ¶ 61,069 (25 Jan. 1988) refers to volume 42 of *FERC Reports,* paragraph 61,069, issued on the parenthetical date.

U.S., F. or *F.2d,* and *F. Supp.* contain the opinions, respectively, of the U.S. Supreme Court, the U.S. courts of appeals, and the U.S. district courts. These volumes are organized chronologically. Any citation includes the relevant volume and page numbers and date of the opinion. A citation to *U.S.* means, by definition, that the opinion was issued by the U.S. Supreme Court; a citation to *F.2d* or *F. Supp.* includes the name of the appeals or district court, respectively, that issued the opinion. For example, 510 F.2d 198 (D.C. Cir. 1975) refers to volume 510 of the *F.2d Reports* at p. 198, decided by the U.S. Court of Appeals for the District of Columbia in 1975.

The public libraries in some large cities carry the *U.S. Code, Code of Federal Regulations,* and possibly some of the other materials discussed here. Check with your local librarian. A better bet—particularly for locating the more technical documents, such as the *Federal Register* and *FERC Reports*—is a law school library. If you live near a law school, contact its librarian about access to those documents. A third possibility is to enlist an attorney, either as a volunteer or for a fee; a law firm with its own library, or with access to computerized data bases such as LEXIS and Westlaw, will be able to locate legal documents relevant to your intervention motion.

Glossary of Hydropower Terms

Acre-foot. The volume of water that would cover one acre to a depth of one foot.

Cfs. Cubic feet per second, a standard measure of the total amount of water passing by a particular point along a river.

Conduit project. A project that utilizes the hydroelectric generating potential of a constructed conduit, such as an irrigation or water supply canal.

Diversion. A structure that diverts water out of a stream or river into a penstock or canal to generate hydropower. Unlike a dam, a diversion structure often does not block the entire river.

Exemption. A form of FERC authorization to construct, available for (1) projects with a capacity of less than five megawatts that utilize preexisting dams or natural water features and (2) projects that utilize the hydroelectric generating potential of constructed conduits, such as irrigation canals.

Forebay. The body of water immediately upstream from the generating plant, from which water is fed into the generators.

Head. The difference in elevation between the surface of the water before it enters the plant (forebay) and the surface of the water below a hydroelectric project (tailrace).

Headwater benefits. The electricity-generating benefits resulting from the storage or release of water upstream from a hydropower generator.

Installed capacity. The total capacity of the generating units in a hydroelectric plant, as listed on the plant equipment.

Kilowatt (kw). A measure of electric power or generating capacity, equal to 1,000 watts or 1.34 horsepower.

Kilowatt-hour (kwh). A measure of electric energy, equal to the quantity of electric energy produced by 1,000 watts operating for one hour. Electric energy is sold by the kilowatt-hour.

Levelized cost. The actual cost of an investment expressed in equal periodic payments over a specified period of years.

License. The basic form of FERC authorization to construct a new project or continue operating an existing project. Every operating hydroelectric project subject to Commission jurisdiction must have either a license or an exemption.

Load. The total amount of electric energy generated by a plant. The average load on a plant is calculated from the total energy it generates over a year divided by the number of hours in a year (8,760).

Megawatt. A measure of electric power or generating capacity, equal to 1,000 kilowatts.

Peak load plant. A power plant that is operated to provide energy only during maximum demand periods.

Penstock. A pipe or tube that transports water from a dam or point of diversion to a hydroelectric generator.

Plant factor. The ratio of the average load on a plant to the plant's installed capacity.

Power pool. Two or more interconnected electric systems which operate as a single system to supply power to meet combined load requirements.

Preliminary permit. A FERC authorization granting priority right to file a license application and authorizing the permittee to conduct studies and analyses necessary to prepare a complete license application. A preliminary permit does not permit any construction.

Pumped-storage project. A hydroelectric plant that generates energy for periods of peak demand or emergency demand by using pumped storage, that is, water pumped previously, during off-peak periods, from a lower-elevation reservoir to a higher-elevation reservoir.

Relicensing. The administrative proceeding in which the Commission, in consultation with other federal and state agencies, decides whether and on what terms to issue a new license for an existing hydroelectric project at the expiration of the original license.

Reversible turbine. A hydraulic turbine, normally installed in a pumped-storage plant, which can be used alternatively to generate electricity or to pump water to a higher elevation.

Run-of-river. A type of hydroelectric project in which the amount of electricity generated is controlled mainly by the volume of water flowing in the stream above the project. Any project which cannot store significant quantities of water at or above the site must be operated as a run-of-river facility.

Service area. Territory in which a utility system or distributor provides service to consumers.

Tailrace. The body of water into which water flows after passing through the generating turbines.

Wheeling charge. An amount paid by a power producer for the right to transmit energy over another company's transmission lines.

Index

A

Acid rain, 3
Adjudicatory (evidentiary) hearings, 40, 71, 87, 99
Administrative Procedure Act, 92, 99
Allies in intervention, 97–98, 125–26
American Canoe Association, 23
American River, 6, 88–89, 90
American Rivers, 11, 12, 18, 46, 114, 125, 127, 128, 129
American Whitewater Affiliation, 11, 23, 125, 127, 128
  Hydromania Award, 129, 130–31
Appeals, 89–92
  of decision by FERC director, 90–91
  of decision by full Commission, 91–92
Application review of process, *see* Federal Energy Regulatory Commission, application review process
Army Corps of Engineers, U.S., 1, 18
  jurisdiction of, 16, 117
  Section 404 discharge permits issued by, 117–18
Attorneys, assistance from, 85, 126–27

B

Bass, 101
Beach erosion, 5
Big A hydroelectric project, 46–48
Bonneville Power Administration, 121
Bull Bridge Gorge, 22–23
Bureau of Land Management (BLM), 1, 51, 52, 55, 97, 106, 107, 113, 118, 119
  consultation of license applicants with, 36
Bureau of Reclamation, 18

C

California Save Our Streams, 89
Cataract Project, 18, 32

Categorical exclusion, 57
Certificate of service, 22
Citing legal authorities, 98
Clean Water Act, 121
  1972 amendments to, 16
  Section 404 discharge permits, 117–18
  Section 401, 36, 47, 116–17, 119
  state water quality certifications under, 116–17
Cold-water fisheries, 8
Columbia Falls project, 6–7
Columbia River Gorge National Scenic Area, 116
Comment date, 34, 71, 84
Comprehensive planning requirement, 48–51, 66–67, 100–101, 119–20
  FERC plans, 49–50, 100–101
  plans prepared by others, 50–51, 101
Confederated Tribes and Bands of the Yakima Indian Nation v. FERC, 75, 100, 105–106
Connecticut Light and Power, 23
Conservation groups, national, 125–26, 127, 128
  addresses and telephone numbers of, 134
  see also names of individual groups
Conservation Law Foundation, 47
Consultation of FERC with other entities, 52–56, 119–20
  federal land management agencies, 54
  fish and wildlife agencies, 53
  miscellaneous requirements, 55–56
Consultation of license applicants with other agencies, 36–37
  nonpower license, 76
  public participation in, 71
  during relicensing, 70–71
Context, environmental, 106
Coordinating your position with allies, 97–98
Council on Environmental Quality (CEQ), 57, 59, 106
Court of Appeals for the Ninth Circuit, U.S., 49–50, 75, 89, 108
Cumulative environmental impacts, 58–59
Cumulative impacts analysis, motion for, 87

D

Dams and hydropower:
  energy from "small hydro" sites, 9, 10, 24
  legislation affecting construction of, *see specific legislation, e.g.* Electric Consumers Protection Act; Federal Power Act; Public Utility Regulatory Policies Act of 1978
  licensing of, *see* Federal Energy Regulatory Commission, original licensing by
  number of existing, 1–2
  proposals for new, 2, 7–8
    see also Relicensing
  relicensing of, *see* Federal Energy Regulatory Commission, relicensing by
  unlicensed, 18–19
  ways in which rivers are harmed by, 4–7
    illustrations of, 6–7
    low oxygen water, 4
    sedimentation, 5
    streamside ecology, 4–5
    temperature stratification, 4
  see also individual dams
Deferral of material issues until after licensing, 99–100
Delaware River, 112
Denial of a new license, 72–73
Department of Agriculture, 55
Department of Energy, U.S., 19
Department of the Interior, 55
Direct mail, fund-raising by, 129
Discovery at adjudicatory hearing, 40
Dispute resolution, methods for, 99–100
Docket sheet, 81
Douglas, William O., 45

E

Edwards Dam, 12
Electric Consumers Protection Act (ECPA), 9, 24, 25, 33, 108, 116
  provisions of, 17, 45, 46
    comprehensive planning requirement, 48, 50, 101, 119–20
    Section 8(d), 59–61
  relicensing and, 67–68, 69, 71
Energy conservation, 46, 52
Environmental analysis, 118
  by FERC, 37–40, 57
Environmental assessment (EA), 37, 57–58, 71
Environmental impacts, FERC's responsibility to study, 56–59, 71
  consideration of alternatives, 59
  cumulative environmental impacts, 58–59
  level of NEPA analysis, 57–58
Environmental Protection Agency (EPA), 1, 97, 117
  consultation of license applicants with, 36
Environmental quality, protection of, 46
"Equal consideration" test, 45–46, 67
Escondido Mutual Water Co. v. LaJolla Band of Mission Indians, 55
Evidentiary hearings, 40, 71, 87, 99
Exemptions, seeking, to Federal Power Act licensing requirements, 29, 31–32
Exhibit to license application explaining compliance with consultation requirements, 37
Experts, assistance from, 126–28
External costs, 108

F

Factual evidence, presenting, 96–97
Federal Energy Regulatory Commission (FERC), 6
  addresses and telephone numbers, 133
  adjudicatory hearing, 40, 71, 87, 99
  administrative law judges at, 21
  application review process of, 33–40, 71
    acceptance for filing, 34
    consultation with other agencies, 36–37
    environmental analysis, 37–40, 71
    information required in applications, 34, 35
    notice in the *Federal Register* and newspapers, 34–36, 71
    project number, 33
  commission decision, 40
  communications with employees of, 88
  courtlike character of, 21–22, 96
  exemptions, 29, 31–32
  Freedom of Information Act regulations of, 169–82
  guidelines on appropriate level of NEPA analysis, 58
  history of, 7–8, 16–17
  jurisdiction of, 7, 9, 16, 17–19, 24
  legal documents on which FERC relies, finding, 197–98
  licensing order, example of, 148–58
  NEPA regulations of, 105, 106, 159–60
  Office of General Counsel, 19
  Office of Hydropower Licensing, 19, 21, 52, 80, 81
    appealing a decision of director of, 90–91
    Division of Dam Safety and Inspections, 19
    Division of Project Compliance and Administration, 19, 93
    Division of Project Review, 19
  opinion of law judge, 40
  organization of, 19–21
  original licensing by, 6, 28–32
    licenses, 29–30
    preliminary permits, 28–29, 35–36

Federal Energy Regulatory Commission (FERC) (*continued*)
standards and procedures for, 44–61
other responsibilities of, 19
overview of, chart, 38–39
participating in FERC process, 80–93
intervening in a FERC proceeding, 83–89, 135–47
learning about a project, 80–83
rectifying a FERC, 89–92
vigilance after project construction, 92–93
prodevelopment policy of, 8
project managers, 21
public information room, 33, 81, 82, 86
raising issues before, 96–108
general rules for making your case, 96 98
potential issues to raise, 98–108
regional offices of, 19
relicensing by, 6, 11–12, 32–33, 64–76
application review process, 33–40
1989–1999 workload, 183–96
procedural aspects for intervening in procedures for, 70–71
standards for, 67–69
service list, 81
standards and procedures for new projects, 44–61
for relicensing, 67–71
Federal land management agencies, 54, 118–19
see also individual agencies, e. g. Bureau of Land Management; Forest Service, U.S.
Federal Land Policy and Management Act (FLPMA), 55, 107, 118
Federal Power Act, 7, 11, 16, 17, 32, 64, 89, 101, 107, 117
on nonpower license issuance, 75
protection of fish and wildlife, 46
"public interest" standard in, 44–45, 101, 117
Section 15, 68
Section 4(e), 51–52, 55, 67, 118
Section 10, 48, 49, 50, 51, 52, 56, 67
Section 10(j) process, 53–54, 120
Section 23b, 72
state water rights and, 120–21
see also Electric Consumers Protection Act
Federal Power Commission, *see* Federal Energy Regulatory Commission
Federal Register, 34, 70, 71, 84
Federal takeover, 33, 73–74
Federal Water Power Act, *see* Federal Power Act
Festivals, fund-raising, 128–29
Financial feasibility of a project, 108
First Iowa Hydro-Electric Cooperative v. FPC, 121
Fish and fishing, 31, 101–103, 117
dams' effects on habitat of, 4, 5
fish passage facilities, 67
protection of, 46, 67, 68
Section 10(j) process, 53–54, 69
relicensing and, 65, 66, 67
see also individual species of fish
Fish and wildlife agencies, state, 31–32, 51, 53, 54, 97, 120
consultation of license applicants with, 36
FERC consultation with, 120
Fish and Wildlife Service, 6, 9, 32, 51, 53, 54, 55, 97
Floods, 2–3
Forest Service, U.S., 51, 52, 55, 97, 106, 107, 116, 118, 119
consultation of license applicants with, 36
wild and scenic river system and, 112, 113
Freedom of Information Act (FOIA), 82–83
Freedom of information request, 82–83
FERC's regulations for, 169–82
Friends of the Earth, 125
Fund-raising, 128–29

G

Gauley River, 126
Gauley River Festival, 128
GEM Irrigation District Power Project, 130
Genesee River, 115
Glossary of hydropower terms, 199–201
Great Northern Paper Company, 46–48
"Gross rivers product," 7

H

Hawks Nest project, 6
Hearings, adjudicatory (evidentiary), 40, 71, 87, 99
Henry's Fork, Salmon River, 116
Hetch Hetchy Dam, 72
Historical preservation office, state, 36
"Home" rivers, 8
Housatonic Area Canoe and Kayak Squad (HACKS), 23
Housatonic River, 22–23
Hydromania award, 129, 130–31
Hydropower, *see* Dams and hydropower
Hydropower coordinator, state, 81

I

Idaho, 121
Idaho Power Company, 108
Idaho Power Co. v. FERC, 108
Indian tribes, 67
  FERC consultation with, 55–56
  see also Reservations, projects on federal, 51–52, 106, 118
In Re Rainsong Co., 106
Intensity of environmental impact, 106
Intervening in a FERC proceeding, 83–89
  getting your motion to FERC, 86
    model motion to intervene, 135–47
  motions and timing, 83–85
  next steps, 87–88
  what the intervention should contain, 85–86
Iowa, 121
Izaak Walton League, 125

J

Jackson, New Hampshire, 114–15
Jackson Falls, 114
Judicial review, 83, 90

K

Keating, Joseph, 88, 89
Kennebec River, 12, 33
King, Dawn, 88
Kingsley Dam, 73
Kootenai Falls, 102
Kootenai Indians, 102
Kootenai River, 102
Kottcamp, Glen, 89

L

Lac Court Oreilles Band of Lake Superior Chippewa Indians v. FPC, 72
LaFlamme, Harriet, 6, 88–89
LaFlamme v. FERC, 49–50, 89, 100
Land management plans, consistency with federal, 106–107
Lawyers, assistance from, 85, 126–27
Learning about a hydroelectric project, 80–83
  FERC records, getting, 81–83
    using Freedom of Information Act, 82–83, 169–82
  getting the project number, 80
  getting on service list, 81
  talking with project manager, 81
  talking with state hydropower coordinator, 81
Legal authorities, citing, 98
Legal documents on which FERC relies, finding, 197–98

Legislative veto of hydroelectric development, 116
Lena Creek, 106
Licenses:
nonpower, 33, 67, 72, 75–76
project license, *see* Project license
relicensing, *see* Relicensing
Lycoming Creek, Pennsylvania, 65

M

McGraw, Chuck, 103
Maine Audubon Society, 47
Maine Board of Environmental Protection, 47
Maine Land Use Regulatory Commission (LURC), 47
Michigan Department of Natural Resources, 33
Model motion to intervene, 135–47
Monongahela Power Co. v. Marsh, 117
Montana, 102, 121
Motion for cumulative impacts analysis, 87
Motion for leave to file a late intervention motion, 22, 84–85
Motion to intervene, 22, 34, 71, 93
timing of filing, 83–84
what it should contain, 85–86
filing, 86
model motion to intervene, 135–47

N

Namekagon River, 101–102
National Audubon Society, 125
National Environmental Policy Act (NEPA), 37, 56–59, 66, 68, 89, 117, 118
FERC's guidelines on appropriate level of NEPA analysis, 58
FERC's NEPA regulations, 105, 106, 159–68
National Marine Fisheries Service, 51, 53, 54, 97
National Park Service, 1, 51, 97, 112, 119
wild and scenic rivers system and, 112, 114
National Resources Council of Maine, 47
National scenic area designation, 116
National Wildlife Federation, 125
National Wildlife Refuge, 1
Nationwide Rivers Inventory, 51
Natural resources, proving impacts on, 101–105
"Natural water features" exemption, 31
Net investment, compensation of licensee for, 73–74
New River, 6
Nonpower license, 33, 67, 72, 75–76
procedure for obtaining, 75–76
Northwest Power Planning Council, 121–22
Notice of intent to seek relicensing, 70

O

Ocoee River Festival, 128
Olson, W. Kent, 1–13
Olympic National Forest, 106
Oregon, 116, 121
Owens River basin, 102
Oxygen in water, 117
dams' effects on, 4, 47

P

Pacific Northwest Electric Power Planning and Conservation Act, 121
Parks department, state, 97
Payette River, 130, 131
PCBs, 23
Pemigewasset River, 115
Penobscot River, 46–48, 128
Permits, preliminary, *see* Preliminary permits
Petition for appeal, 90
Petition for rehearing, 91–92
Pine River, 66

Platte River, 73
Pleasant River, 6–7
Politicians, seeking allies in, 126
Pollution control agency, state, 97
Postlicensing stage, vigilance in the, 92–93
Power sale contracts, 74
Preliminary permits, 28–29, 84
competing, 28–29
purpose of, 28
requirement for obtaining, 28
terms of, 29
Preservation of natural rivers, reasons for, 2–4
Press coverage, gaining favorable, 129–31
Procedural issues, raising, before FERC, 97
Procedures, *see* Standards and procedures of Federal Energy Regulatory Commission
Project costs, 108
Project license, 29–30
amending the, 30
categories of applications for, 30
commencing of construction, period for, 30
competing applicants for a, 29
conditions detailed in a, 29
information required to receive a, 29–30
reopener clauses in, 30
special conditions, 98
term of, 30, 32, 64
Project manager, FERC, 81
Project number, FERC, 80
Protected rivers, 9, 60–61
state, 119–21
statistics on, 1–2
under Wild and Scenic Rivers Act, 112
"Public interest" standard, 44–45, 67, 72, 96, 101, 117
Public Utility Regulatory Policies Act of 1978 (PURPA), 31, 44, 50, 107–108
environmental standards for PURPA benefits, 59–61
intent of, 24–25
purpose of, 8, 24–25
state veto of PURPA benefits, 120
Pyramid Creek, 88

Q

Quality of life, preservation of rivers to improve, 3

R

Raffles, fund-raising, 129
Rainbow trout, 102
Reagan, Ronald, 115
Recreational use of rivers, 2, 10
nonpower license and, *see* Nonpower license
protection of, 46
Recreation plans, state, 51
Rectifying a FERC error, 89–92
appealing a decision by the director, 90–91
appealing a decision of the full commission, 91–92
obtaining a stay of an order, 92
Relicensing, 6, 11–12, 32–33
application review process, 33–34
denial of new license, 72–73
incumbent licensee and, 68–69
issues raised in, versus those in original licensing, 64–66
by issuing new license, 33, 71
nonpower license, 33, 67, 72, 75–76
overview, 64–67
procedural aspects of intervening in procedures for, 170–71
recommendation of a federal takeover, 33, 73–74
standards for, 67–69
Reservations, projects on federal, 51–52, 106, 118
Rocky Mountain Power Company, 107–108

Rules for making your case before FERC, 96–98
  cite legal authorities, 98
  encourage your allies to make similar arguments, 97–98
  focus on procedural as well as substantive issues, 97
  support your argument with as many facts as possible, 96–97

S

Saco River, 18, 32, 114
Salmon:
  Columbia Falls project and, 6–7
  Penobscot River, 47
  Saco River, 18
Savage River, 103–105, 112, 116
Savage River Defense Fund, 129
Sayles Flat, 6, 88–89, 90
Scenic river programs, state, 50–51
Scoping session, 40
Secretary of the Interior, 113
Service list, FERC, 81
Severance damages, 74
Sierra Club, 126
Society for New Hampshire Forests, 114–15
Soil Conservation Service, U.S., 1
Standards and procedures of Federal Energy Regulatory Commission:
  importance of experts in dealing with, 126–28
  for new projects, 44–61
    comprehensive planning requirement, 48–51, 101
    environmental standards for PURPA benefits, 59–61
    "equal consideration" standard, 45–46
    miscellaneous licensing requirements, 51–52
    "public interest" standard, 44–45, 96, 101, 117
    responsibility to consult with other entities, 52–56
    responsibility to study environmental impacts, 56–59
  for relicensing, 67–71
    consulting with other agencies, 70–71
    deadlines for filing application for, 70
State-designated wild and scenic river, 113
State programs and agencies, *see specific types of programs and agencies, e.g.* Fish and wildlife agencies, state; Scenic river programs, state
State river protection, 119–21
State water rights, 120–21
Stay of an order, obtaining a, 92
Stay of effective date of a new license, 74
Steamboaters v. FERC, 105
Strategic planning, 124
Strategies for effectively dealing with hydroelectric projects, 124–31
  favorable press coverage, 129–30
  finding allies, 125–26
  legal and technical expert assistance, 126–28
  raising money, 128–29
  strategic planning, 124
Supreme Court, U.S., 45, 55, 121
Swan Falls Dam, 18–19

T

Takeover by the government, 33, 73–74
Technical assistance, 127–28
Telephone, fund-raising by, 129
Temperature, 117
  stratification of, in rivers, 4
Thornton, Mac, 105
Trout, 4, 102, 103

Trout Unlimited, 126
Tulalip Tribes v. FERC, 31
Tuolumne River, 112, 128

U

Udall v. FPC, 45
United Mine Workers, 125

V

Veto of hydroelectric development, legislative, 116
Vision questing, 102

W

Wall Street Journal, 1
Washington, 116, 121
Water quality certifications, state, 116–17
Water resource plans, state, 51
Western Power, 130
White-water, adventure-class, 8, 64, 117
  intervention to protect, 22–23, 47–48, 103–105
Whooping cranes, 73
Wild and Scenic Rivers Act, 50, 112, 116
  Section 2(a)(ii), 113
Wild and scenic rivers systems, 112–15, 125
  categories of rivers, 113–14
  qualifying characteristics, 113–14
  ways rivers can be added to, 112–13
  Wildcat Brook case, 114–15
Wildcat Brook, 114–15
Wildcat River Trust, 115
Wildlife, 31–32, 101–103
  protection of, 46, 67, 68
    Section 10(j) process, 53–54, 69
  see also specific types of wildlife
Wildlife agencies, state, *see* Fish and wildlife agencies, state
Wilmer Cutler and Pickering, 47
Wilson, Ron, 103
World Canoe and Kayak Championship, 103–105

Y

Yosemite National Park, 72

About the Authors

W. KENT OLSON. Mr. Olson, the author of the Introduction to *Rivers at Risk,* has been the president of American Rivers since January 1986. Prior to joining American Rivers, Mr. Olson served as executive director of the Connecticut chapter of the Nature Conservancy, publisher and editor in chief of the Appalachian Mountain Club's guidebooks and its journal *Appalachia,* and director of the Appalachian Mountain Club hut system. Mr. Olson has held faculty appointments at Wesleyan University and Yale College, and at Yale School of Forestry and Environmental Studies, from which he earned his master's degree in natural resource management.

JOHN D. ECHEVERRIA. Mr. Echeverria is the general counsel of American Rivers and the director of its National Center for Hydropower Policy. Prior to joining American Rivers in 1987, he was an associate with the law firm of Hughes Hubbard and Reed in Washington, D.C., and law clerk to the Honorable Gerhard Gesell. Mr. Echeverria is a graduate of the Yale Law School and the Yale School of Forestry and Environmental Studies. A fisherman and canoeist, Mr. Echeverria has written extensively on energy and environmental matters.

POPE BARROW. Mr. Barrow is an attorney with substantial experience in environmental and energy law. He is a director of the American Whitewater Affiliation. Mr. Barrow, an avid white-water enthusiast, has written extensively on river conservation issues. In 1987, in recognition of his exceptional contributions to river conservation, Perception, Inc., awarded Mr. Barrow the River Conservationist of the Year Award. Mr. Barrow is a graduate of Yale College and Harvard Law School.

RICHARD ROOS-COLLINS. Mr. Roos-Collins is a Deputy Attorney General in environmental law for the state of California. Before taking his present position, he was a staff attorney with the Office of General Counsel of the U.S. Environmental Protection Agency in Washington,

D.C. Mr. Roos-Collins was formerly publicist and litigation coordinator for Friends of the River, a San Francisco–based nonprofit organization dedicated to preserving California's wild rivers. In that capacity, Mr. Roos-Collins prepared a layperson's guide to intervening in FERC proceedings, which provided the foundation for much of the material in *Rivers at Risk.* He is a graduate of Princeton and of Harvard Law School.

Also Available from Island Press

The Challenge of Global Warming
Edited by Dean Edwin Abrahamson
Introduction by Senator Timothy E. Wirth
In cooperation with the Natural Resources Defense Council
1989, 355 pp., tables, graphs, bibliography, index
Cloth: $34.95 ISBN: 0-933280-87-4
Paper: $19.95 ISBN: 0-933280-86-6

The Complete Guide to Environmental Careers
By The CEIP Fund
1989, 355 pp., photographs, case studies, bibliography, index
Cloth: $24.95 ISBN: 0-933280-85-8
Paper: $14.95 ISBN: 0-933280-84-X

Down by the River: The Impact of Federal Water Projects and Policies on Biological Diversity
By Constance E. Hunt with Verne Huser
In cooperation with the National Wildlife Federation
1988, 256 pp., illustrations, glossary, bibliography, index
Cloth: $34.95 ISBN: 0-993280-48-3
Paper: $22.95 ISBN: 0-993280-47-5

Natural Resources for the 21st Century
Edited by R. Neil Sampson and Dwight Hair
In cooperation with the American Forestry Association
1989, 350 pp., illustrations, index
Cloth: $34.95 ISBN: 1-55963-003-5
Paper: $19.95 ISBN: 1-55963-002-7

The Poisoned Well: New Strategies for Groundwater Protection
Edited by Eric P. Jorgensen
In cooperation with the Sierra Club Legal Defense Fund
1989, 400 pp., glossary, charts, appendixes, bibliography, index
Cloth: $31.95 ISBN: 0-933280-56-4
Paper: $19.95 ISBN: 0-933280-55-6

Reopening the Western Frontier
Edited by Ed Marston
From *High Country News*
1989, 350 pp., illustrations, photographs, maps, index
Cloth: $24.95 ISBN: 1-55963-011-6
Paper: $15.95 ISBN: 1-55963-010-8

Research Priorities for Conservation Biology
ISLAND PRESS CRITICAL ISSUES SERIES
Edited by Michael E. Soulé and Kathryn A. Kohm
In cooperation with the Society for Conservation Biology
1989, 110 pp., photographs, charts, graphs
Paper: $9.95 ISBN: 0-933280-99-8

Rush to Burn
From *Newsday*
Winner of the Worth Bingham Award
1989, 276 pp., illustrations, photographs, graphs, index
Cloth: $22.95 ISBN: 1-55963-001-9
Paper: $14.95 ISBN: 1-55963-000-0

Shading Our Cities: Resource Guide for Urban and Community Forests
Edited by Gary Moll and Sara Ebenreck
In cooperation with the American Forestry Association
1989, 260 pp., illustrations, photographs, appendixes, index
Cloth: $34.95 ISBN 0-933280-96-3
Paper: $19.95 ISBN: 0-933280-95-5

The Sierra Nevada: A Mountain Journey
By Tim Palmer
1988, 352 pp., illustrations, appendixes, index
Cloth: $31.95 ISBN: 0-933280-54-8
Paper: $14.95 ISBN: 0-933280-53-X

War on Waste: Can America Win Its Battle with Garbage?
By Louis Blumberg and Robert Gottlieb
1989, 325 pp., charts, graphs, index
Cloth: $34.95 ISBN: 0-933280-92-0
Paper: $19.95 ISBN: 0-933280-91-2

Wildlife of the Florida Keys: A Natural History
By James D. Lazell, Jr.
1989, 254 pp., illustrations, photographs, line drawings, maps, index
Cloth: $31.95 ISBN: 0-933280-98-X
Paper: $19.95 ISBN: 0-933280-97-1

These titles are available from Island Press, Box 7, Covelo, CA 95428. Please enclose $2.00 shipping and handling for the first book and $1.00 for each additional book. California and Washington, DC, residents add 6% sales tax. A catalog of current and forthcoming titles is available free of charge.